D0845750

A Passion for Gold

A Passion for Gold

AN AUTOBIOGRAPHY

Ralph J. Roberts, Ph.D.

With Mary Beth Gentry

University of Nevada Press
Reno & Las Vegas

University of Nevada Press, Reno, Nevada 89557 USA

Manufactured in the United States of America
Design by Kaelin Chappell

Library of Congress Cataloging-in-Publication Data
Roberts, Ralph Jackson, 1911–
A passion for gold : an autobiography / Ralph J. Roberts
with MaryBeth Gentry.
p. cm.
Includes bibliographical references and index.
ISBN 0-87417-502-X (hardcover : alk. paper)
1. Roberts, Ralph Jackson, 1911–
2. Geologists—United States—Biography.
I. Gentry, Mary Beth, 1945– II. Title.
QE22.R59 A3 2002
551'.092—dc21
2002004779

The paper used in this book meets the requirements of
American National Standard for Information Sciences—
Permanence of Paper for Printed Library Materials, ANSI Z39.48-1984.
Binding materials were selected for strength and durability.

Frontispiece: Ralph J. Roberts, May 1998. Photograph by Ted Cook,
University of Nevada, Reno, reprinted with permission.

Portions of the *American Geological Institute Glossary of Geology*
are reprinted with permission.

11 10 09 08 07 06 05 04 03
5 4 3 2

To Arleda, Michael, Steven, Kim, and Mij

Contents

Illustrations

FIGURES

MAPS

Preface

What's past is prologue.

—SHAKESPEARE, *The Tempest*

A boat, beneath a sunny sky,
Lingering onward dreamily
In an evening of July—

Children three that nestle near,
Eager eye and willing ear,
Pleased a simple tale to hear—

Long has paled that sunny sky:
Echoes fade and memories die:
Autumn frosts have slain July

Still she haunts me, phantomwise,
Alice moving under skies
Never seen by waking eyes.

Children yet, the tale to hear,
Eager eye and willing ear,
Lovingly shall nestle near.
In a Wonderland they lie,
Dreaming as the days go by,
Dreaming as the summers die:

Ever drifting down the stream—
Lingering in the golden gleam—
Life, what is it but a dream?

—LEWIS CARROLL,
Through the Looking Glass

Although Lewis Carroll's *Through the Looking Glass* was written as children's fantasy, it fits nicely into the study of geology. Geology is not a precise science but requires a great deal of imagination and ingenuity. So, I have used quotes from Carroll's books, as well as from his poem, above, to emphasize how we must occasionally go beyond our conventional ways of thinking and look at all sides of a question. As the White Queen told Alice, "sometimes I've believed as many as six impossible things before breakfast." With flexibility of mind, geologists can solve any problems they may encounter. Such flexibility was required in the search for gold in Nevada.

Gold is more than a gleam to me, for the exquisite beauty of its natural crystalline form has led to its being prized by both ancient and modern man. Art objects fashioned of everlasting gold grace many great museums today.

Looking back, I find that my life seems like a dream. In the beginning I had no firm objectives. I just wanted to live a quiet life, but a series of events carried me from one challenging adventure to another. As Carroll said, "Echoes fade and memories may die," but my memories will live forever. Life has blessed me not once, but four times. I have been favored with four great loves in my life. At the age of twenty-eight, I met my first love, Arleda, my soul mate for forty-six years. Second, we were blessed with three wonderful sons. Arleda passed away in 1988. Then four years later, at the age of eighty-one, I rediscovered my third and present love, Mij, whom I had known in high school. My fourth love was my life's work, geology, which I first embraced in college; I found a niche in the U.S. Geological Survey (USGS) and enjoyed a wonderful career there. With my mentor, Henry Ferguson, and others, I pursued a life of geologic investigation, and we ultimately played major roles in several significant discoveries. I attribute much of my good fortune to being in the right place at the right time, but when inspiration tapped me on the shoulder, I eagerly responded "full speed ahead, and damn the consequences."

This is especially true of the geological discoveries that have pointed the way to a recognition of the Antler Orogeny, which led to the correct interpretation of the geological framework of Nevada. Understanding the framework permitted us to help mining geologists discover gold deposits of large size and fabulous richness. We (Roberts and others, 1967) recognized these large deposits as belonging to a new type and called them Carlin-type deposits. They are low-temperature deposits that replace limestone in fractured and faulted zones. These finds led to the renewal of gold mining in Nevada, which had begun on a small scale in the 1930s with the discovery of gold at the Getchell and Gold Acres Mines. They will continue to be productive well into the twenty-first century. They will have yields far exceeding the bonanza deposits discovered in the late 1800s. Wrapped up in the discoveries of Carlin-type deposits is the essence of my life's work—over forty-four years of traversing sagebrush- and juniper-covered mountains and valleys, mapping and assembling the detailed structural and stratigraphic information into a coherent geologic framework.

Many geologists preceding Henry Ferguson and me noted significant and useful facts about the mines and geology of Nevada, but they did not tie these facts together. Expanding on Ferguson's 1924 and 1929 work in central Nevada, I was able in 1949 to describe a major mountain-building event (the Antler Orogeny) of late Paleozoic age in north-central Nevada. In 1962, Norman Silberling and I defined a later orogeny, the Sonoma. These orogenies were major factors in the upward movement of gold-, silver-, and copper-bearing solutions during Tertiary time.

It was my good fortune to work in Eureka County, Nevada, with Robert Lehner during the mid-1950s. Eureka County proved to be the Rosetta stone of north-central Nevada; what we learned from there permitted us to tie the geology and ore deposits of central Nevada into the broader geologic framework of the state.

The deposit at Gold Acres in central Eureka County was the key to the great gold deposits in the Carlin area. In 1955 Lehner and I plotted the region's ore deposits and found that most of them lay along northwesterly trends that defined clear-cut mineral belts. Once these belts were recognized, Newmont Mining Company's chief geologist, Robert Fulton, supervised prospecting by John Livermore and Alan Coope, who discovered the Carlin No. 1 Mine in 1961. This mine had yielded more than three million ounces of gold by 1980, and its successful operation led to a revival of gold mining in Nevada.

The unique aspect of this discovery was that it was made by geologists, Fulton, Livermore, and Coope, who paid me the ultimate compliment—they read my published work and then meticulously tested my recommendations. USGS geologists cannot ask for more.

Throughout this period, I continued to focus on the concept of mineral belts. I presented this idea in 1957 at a meeting in Reno. By this time, it was clear that the concept was a valid prospecting tool, and though many in the mining industry looked upon it with guarded skepticism, I forged ahead, using it in my work on regional geology and related ore deposits. My later studies (with Edwin Tooker) of the Oquirrh Mountains and Bingham district of Utah would demonstrate that stratigraphic and structural controls of ore deposits are indeed critical.

In 1971 fate led me to Saudi Arabia to assist the Saudis in developing their fledgling minerals program. There my attention was quickly captured by the enormous potential of the Mahd adh Dhahab Mine, which Robert Luce, Ron Worl, and I studied. We are convinced that this mine was the major source of King Solomon's gold.

I returned to the United States in 1978. After retiring in 1981 with time, energy, and ideas to burn, I joined a group of close friends in the consulting business to form Victor E. Kral Associates (VEKA). We discovered a new ore body at the Marigold Mine, later to be worked by the Rayrock Company. We also discovered the REN deposit in the Carlin Belt. After years of helping others find gold, I learned that it could be fun to find it for myself!

Later, I joined Raul Madrid in consulting work on other properties. These included the Goldstrike Mine owned by PanCana and Western States Mining Company in the Carlin Belt. After we helped these companies explore Goldstrike, the property was picked up by American Barrick and is now the top gold producer in Nevada with reserves of about fifty million ounces.

Developing the geologic framework of north-central Nevada has been a long and arduous process, and I am deeply grateful to the trailblazers—Edwin Kirk, Charles Merriam, Charles Anderson—and to my mentor, Henry Ferguson, for providing the signs for me to follow. I am grateful that I have been able to carry on their precepts, thus enabling other geologists to discover new ore deposits in the Carlin, Battle Mountain, and Getchell mineral belts. It is a delight to know that my construction of the geologic puzzle that is north-central Nevada has stood the test of time.

ACKNOWLEDGMENTS

I am indebted to many people who have made valuable contributions to this book. These include colleagues Edwin Tooker, the late Hal Morris, Raul Madrid, Arthur Radtke, Alfred Hofstra, David Brew, Keith Ketner, John Stewart, Forrest Poole, Frank Whitmore, Richard Hose, Donald Hadley, Dwight Schmidt, Robert Luce, Ron Worl, William Overstreet, James Norton, Howard Gould, Victor Kral, the late Robert Reeves, Norman Silberling, Jon Broderick, Robert Kamilli, Thor Kiilsgaard, Conrad Martin, and Abdulaziz Bagdady.

Odin Christensen, Don Hausen, John Livermore, the late Alan Coope, Larry Kornze, and Keith Bettles of major mining companies in the Carlin area also furnished valuable information.

H. E. Zaki Yamani, H. E. Ghazi Sultan, and H. E. Fadil Kabbani of the Ministry of Petroleum and Mineral Resources aided my work in Saudi Arabia. Al Amir Abdul Rahman Assudeiri facilitated my work at Mahd adh Dhahab.

Those who helped assemble this book include Martha Hopwood, Tyler Jones, and Mary Kathleen Hagerman Price.

Archaeologists who read and contributed to the chapter on Ophir include Gus Van Beek, U.S. National Museum; Peter Parr, University of London; Kenneth Kitchen, University of Liverpool; Gary Rollefson, Whitman College; and Aharon Horowitz, Tel Aviv University.

My companion, Mij Ogden, has offered steadfast love, support, and encouragement; in addition, she has helped with research on the manuscript. I am also indebted to my sons, Michael and Kim, for constant help and for their memories of family life.

John Rodgers of Yale University made useful suggestions. Maxwell Morgan, Edward Rowe, and Jan Van't Groenewout helped me to bring a layman's perspective to some geologic discussions.

A Passion for Gold

Chapter 1

Geology 101

NEVADA GOLD AND MOUNTAIN BUILDING

> "The time has come," the Walrus said,
> "To talk of many things, of shoes and
> ships and sealing wax, and of cabbages
> and kings."
>
> —LEWIS CARROLL,
> *Alice's Adventures in Wonderland*

Gold, the metal of exquisite beauty, has intrigued ancient and modern man alike.

If ancient man found crystals of gold, probably in quartz veins that are sometimes exposed in mineralized areas, he might have fashioned primitive jewelry by stringing the crystals on a thong cut from a skin of some animal he had killed and eaten. So adorned, he would have been the envy of his tribe. Now we find gold crystals in mines dug deep into the earth, and they are as prized today as they were in ancient times. The ore is also found as nuggets of placer gold in streams. In 1849 in California and in 1898 in Alaska and the Canadian Yukon, prospectors were excited by discoveries of placer gold.

Gold is uniquely malleable and has long been the metal of choice for fashioning art objects; it can be carved, cast, or beaten into tissue-thin sheets. In 1539 Benvenuto Cellini of Florence, Italy, began to fashion a gold saltcellar for King Francis I of France. This magnificent work was finished in 1543 and is now displayed in the Vienna Art Museum (Wienkunst Museum). I can easily understand Cellini's dou-

FIG. 1.1. *Leaf gold with trigons found in the Wadley Mine, Willow Creek Canyon, Pershing County, Nevada. Courtesy of Lois Calder Baum.*

FIG. 1.2. *Cellini saltcellar. Courtesy of the Vienna Art Museum, Austria.*

ble passion—for gold and for the art made from it. Most comparable works of that period have been melted down, and their artistic value has been destroyed. What a pity!

Gold is found in many geologic environments, as lodes (such as quartz veins found in both cold and hot environments), as bodies that replace (substitute for) favorable beds such as limestone, and as sedimentary placer deposits in stream gravel. The geologic processes that form gold deposits include release of gold from deep in the crust and mantle of the earth, and transportation of it upward, to spots nearer the surface. This can be a slow process, but geologic time is long; there is plenty of time for everything. We will consider events that go back far into Precambrian time, i.e., more than 550 million years ago, as well as events that took place within modern history. A geologic time scale divides geologic history into periods. These periods by and large record events such as sedimentation of related rocks or similar mountain-building events. Hang on tight, we're going on a wild ride!

Even now, at the age of ninety-one and of relatively sound body and mind, I find that geology continues to be a major element in my life. Though I can no longer tackle the rugged mountains of the West, I will continue to study, write, and talk about mountain building and related ore deposits. The western cordillera, or mountains of the western United States, has been my laboratory. The complex processes that formed these mountain ranges over the last 300 million years reveal an earth in slow but constant motion with many geological events occurring together. If we could compress time, the enormity of these changes would overwhelm even the most seasoned geologist.

Orogeny, the geological term for mountain building, has always interested scientists, who early on devised a contraction theory based on a shrinking earth; they compared mountains to wrinkles as on a drying orange. Early geologists knew that the earth was originally molten, and they suggested that as it cooled, it shrank, forming mountains. But had this been the case, mountains would have formed in belts of about the same age. The record showed that mountains had formed at many different times during the earth's long history, so the theory was discarded.

Time Units of the Geologic Time Scale				Development of Plants and Animals
Eon	**Era**	**Period**	**Epoch**	
PHANEROZOIC	CENOZOIC	Quaternary	Holocene	Humans develop
			0.01 Ma	
			Pleistocene	
			1.8 Ma	
		Tertiary	Pliocene	Age of Mammals
			5.3 Ma	
			Miocene	
			23.8 Ma	
			Oligocene	
			35.7 Ma	
			Eocene	
			54.8 Ma	
			Paleocene	Extinction of dinosaurs
			65.0 Ma	and many other species
	MESOZOIC	Cretaceous	Age of Reptiles	First flowering plants
		144 Ma		
		Jurassic		First birds
		206 Ma		
		Triassic		Dinosaurs dominant
		251 Ma		
	PALEOZOIC	Permian	Age of Amphibians	Extinction of trilobites and many other marine animals
		290 Ma		First reptiles
		Carboniferous: Pennsylvanian		Large coal swamps
		323 Ma		
		Carboniferous: Mississippian		Amphibians abundant
		354 Ma		
		Devonian	Age of Fishes	First insect fossils
		418 Ma		Fishes dominant
		Silurian		First land plants
		443 Ma		
		Ordovician	Age of Invertebrates	First fishes
		490 Ma		Trilobites dominant
		Cambrian		First organisms with shells
		540 Ma		
PROTEROZOIC	Collectively called Precambrian, comprises about 87% of the geologic time scale			First multicelled organisms
2500 Ma				
ARCHEAN				First one-celled organisms
				Age of oldest rocks
4000 Ma				Origin of the earth

Numbers on the time scale show millions of years before the present.
Source: Geological Society of America, 1998

FIG. 1.3. *Geologic time scale.*

Much later, geologists began to note that mountains formed mainly on the margins of continents and in subsiding troughs. This concept was called the geosynclinal theory of orogeny. All the great mountain ranges, including the Appalachians and Rockies, formed in basins or troughs, but something was lacking—mountain ranges were generally folded and thrust faulted on a large scale by lateral forces. The nature of these forces was not understood until an explanation was supplied by "plate tectonics" (see Osborne and Tarling, 1996), which is based on the observation that the earth's crust is divided into plates that move relative to one another, at times coming together with a crunch. One cause of this movement is spreading of the sea floor due to periodic intrusion of lava from under the sea floor into oceanic ridges. This forces the plates apart. The North American plate, made up of the western Atlantic Ocean and North America, overrode the eastern margin of the Pacific plate, possibly from late Paleozoic to Cenozoic time, giving rise to a complex sequence of events that include magmatism (melting within the earth) and formation of associated ore deposits in the cordillera.

Establishing a link between orogeny and landscapes is vital to understanding the origin of the varied geological formations of this wonderful country. The landscapes are the result of erosion by water, ice, and wind, which, given enough time, will reduce mountains to mere nubbins.

Mountains are of three principal types:

1. Volcanic
2. Fold
3. Block-fault

We have all seen volcanic mountains on the television news. Their awesome power can be frightening, as exemplified by the eruption of Mount Saint Helens in 1980, and they are an integral part of the western cordillera. Volcanoes in the Pacific Northwest are part of a "Ring of Fire" that nearly encircles the Pacific Ocean. The ring results from the ocean floor being forced beneath the continents, where melting can take place, forming molten magma. The magma may rise through the earth's crust and erupt violently, and the volcanoes act as safety

FIG. 1.4. *Mount Rainier, a volcanic mountain in Washington State, about sixty miles southeast of Seattle. Courtesy of Eulalie Fisher.*

FIG. 1.5. *Fold Mountains, Oquirrh Range, Utah. Courtesy of E. W. Tooker.*

valves, allowing pressure to be released. Mount Rainier is another excellent example of such a volcano, towering 14,410 feet above sea level; it is a composite cone, consisting of alternating lava flows and ash. Kilauea in the Hawaiian Islands is a shield volcano, built up entirely of lava flows.

FIG. 1.6. *Block-fault mountain, Tobin Range, about fifty-five miles south of Winnemucca, Nevada. The white line is a fault scarp ten to fifteen feet high formed during the 1915 earthquake.*

Fold mountains, such as the Alps in Europe and the Appalachians and the Rockies in the United States, including the Oquirrh Range in Utah (as shown in fig. 1.5) were formed at various times by continental collisions and plate movements. The western cordillera also underwent comparable folding during the Antler Orogeny, whose mountains extended north-south through central Nevada in late Paleozoic time. The Sonoma Orogeny followed during latest Paleozoic and earliest Mesozoic (Triassic) times. Orogenies also occurred during later Mesozoic and Tertiary times. Folding in the western cordillera was followed by a breakup of the crust and block faulting.

Block-fault mountains occur in the western cordillera in the Basin and Range Province (see fig. 1.6), the area stretching west from the Rocky Mountains to the Sierra Nevada Range. This region has a complex structural history that began with fold mountains during late Paleozoic and Mesozoic times and culminated in a breakup and collapse of the crust with attendant magmatism. Some blocks sank while others rose, forming a rugged landscape. Ceaseless erosion over millions of years carved the present landscape of steep, fault-bounded mountains separated by flat-floored valleys. In addition, during the breakup, magma from deep within the earth moved upward on faults,

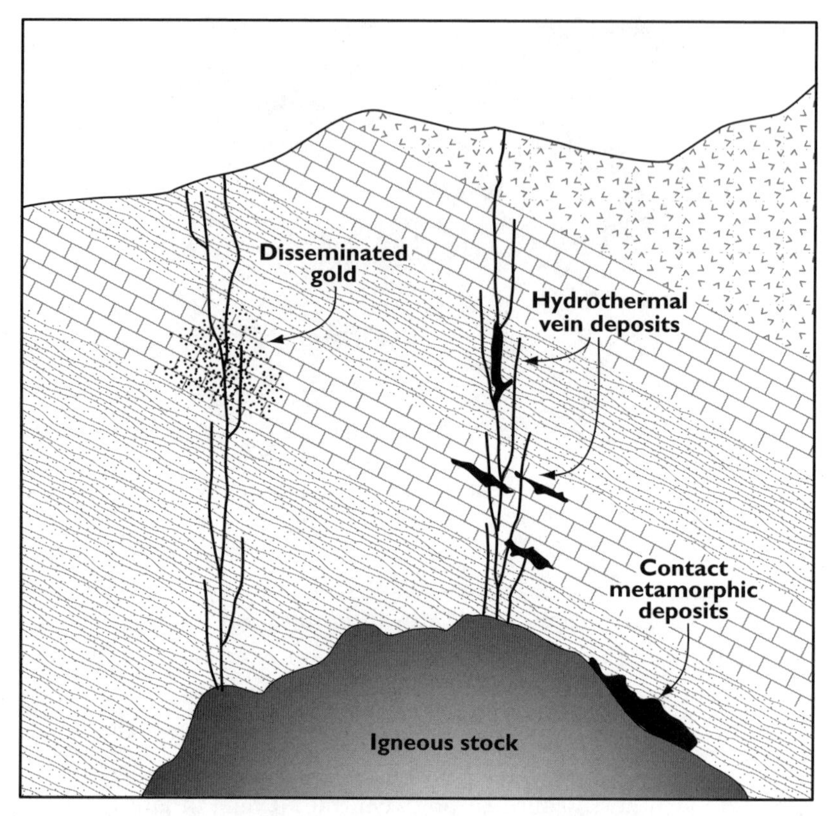

FIG. 1.7. *A typical hydrothermal vein system, showing a granitic source of fluids, the position of contact metamorphic deposits, hydrothermal vein deposits, and disseminated (Carlin-type) deposits.*

forming stocks (small igneous bodies) and dikes (tabular bodies) of granitic rocks. This magmatism was followed by eruptions of volcanic lavas and pyroclastic material such as ash. Then hydrothermal (hot water) solutions carried metals such as gold and silver upward (see fig. 1.7; White, 1955). This fascinating process yielded bonanza deposits of gold and silver.

Gold Deposits

The major gold deposits in Nevada mostly formed in Tertiary time when crustal breakup permitted granitic magma and accompanying gold-bearing hydrothermal solutions to rise into the upper crust. This was a late stage in a long, complex history that began more than 500 million years ago, in Paleozoic time. The process began with sedimentation, mostly of carbonate rocks, such as limestone, along the western continental margin. In late Devonian time, about 360 million years ago, and during early Mississippian time, enormous compressive forces, related to plate tectonics, brought (obducted) oceanic rocks such as chert (a rock like agate), shale, and quartzite and volcanic rocks onto the carbonate shelf rocks of the continent. These movements took place during the first mountain-building phase of the Antler Orogeny, which was closely followed by a second phase that resulted in vertical uplift, forming a chain of mountains. Subsequently, vigorous streams carried coarse debris into basins to the east and the sea to the west.

The orogenic belt gradually eroded to a land of low relief. At the end of Paleozoic or in early Mesozoic time, another thrust plate, the Golconda, moved oceanic shale and chert and volcanic rocks into central Nevada during the Sonoma Orogeny (Silberling and Roberts, 1962). This plate overrode sedimentary rocks previously deposited in basins near the orogenic belt, including the Battle Formation, Antler Peak Limestone, and Edna Mountain Formation, all of which are calcareous, forming another favorable environment for ore deposits.

The Antler and Sonoma Orogenies thus accomplished a very important task: they provided virtually impervious cap rocks for the creation of Carlin-type and Copper Canyon–type gold deposits. The cap rocks prevented or delayed ascending solutions from emerging on the surface during ore formation.

Other major factors in creating these deposits include folding, faulting, and fracturing, and igneous rocks associated with those movements. Regional folding in north-central Nevada involves Jurassic rocks, and is therefore Jurassic or younger (Ferguson, Roberts, and Muller, 1952). This framework established a pattern of northwesterly trending folds and faults in the region, perhaps best shown by the Antler anticline in the vicinity of Antler Peak (Roberts, 1965). Faults and fractures on the flank of the anticline are important features that localize ore at Copper Canyon and Copper Basin. Northwesterly trending folds in the Carlin and Battle Mountain mineral belts are also important ore controls. After the folding, erosion stripped off some of

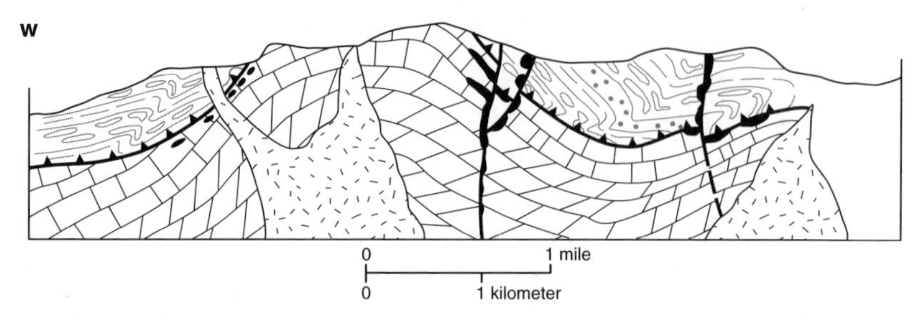

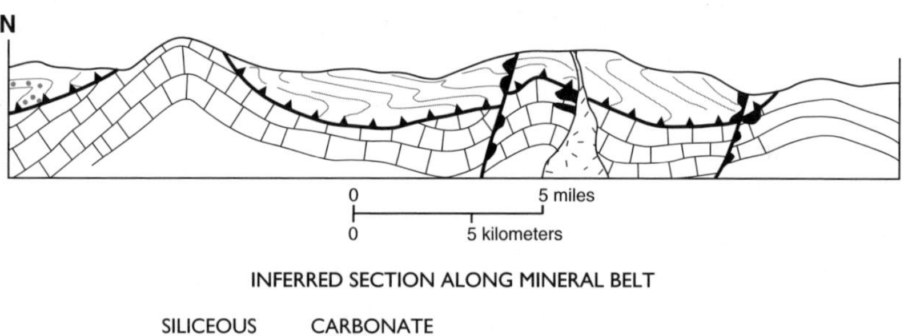

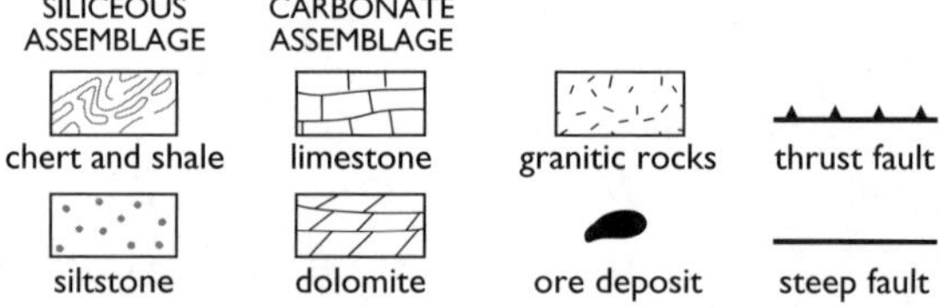

FIG. 1.8. *Schematic showing oceanic rocks formed in deep water, thrust (obducted) onto carbonate rocks of the continental margin. Ore deposits have formed along the thrust zone and in high-angle faults that cut the bedrock.*

the upper plate of the Roberts Mountains thrust, exposing the lower plate rocks in "windows" (see fig. 1.8).

Three principal kinds of gold deposits are found in north-central Nevada: Copper Canyon, Carlin, and veins. Copper Canyon–type gold deposits formed near and within granitic rocks at relatively high temperatures and are characterized by copper, lead, and zinc sulfides. Gold came into the system mostly in late stages of metallization. Carlin-type deposits are large deposits formed at low temperatures mostly in carbonate rocks, such as limestone and dolomite. The principal hosts are Hansen Creek, Roberts Mountains, and Popovich Formation. They are characterized by arsenic, mercury, antimony, and, locally, thallium mineral assemblages. They apparently formed relatively far from magmatic sources but extended to considerable depths in favorable units.

Gold-bearing solutions continued to flow upward in late Tertiary time, but at a reduced rate, forming gold-silver veins. In figure 1.8, the various environments of gold deposits in north-central Nevada are shown.

The heat engines that drove this process were bodies of hot granitic rock, which mainly preceded the gold-bearing solutions. As the solutions ascended, they cooled and reacted with the adjacent wall rocks. Initially, the solutions were slightly acidic; calcareous rocks, such as limestone and dolomite, neutralized the acidic solutions, precipitating the gold and associated metals.

Block faulting and accompanying volcanism occurred through the Cenozoic era and still continue today. Frequent earthquakes in many areas of Nevada provide eloquent testimony that crustal adjustments are still in progress (see fig. 1.6). The principal period for gold formation was from forty-two to thirty-seven million years ago, although both younger and older deposits are known.

This summarizes the essential facts of gold-bearing deposits in Nevada. Now let us turn to the autobiography of Ralph J. Roberts and his family.

Chapter 2

The Early Years

1911–1939

> Always bear in mind that your own resolution to succeed is more important than any one thing.
>
> —ABRAHAM LINCOLN, advice to a young attorney

The Palouse country in southeastern Washington State is known for deep, rich soil that produces abundant crops. The undulating grasslands called the Palouse Hills were sculpted about ten thousand years ago during the Wisconsin glacial age when loess, soil carried by the wind from the outwash of glaciers, covered parts of the Pacific Northwest. Much later, a section of that fertile land was homesteaded by my grandparents, the Hillary Jackson Boozers.

My mother, Rebecca Evangeline (Eva) Boozer, was born in 1885 of German, Swiss, and English stock in New Burnside, Kentucky, and went west in 1887 with her parents on the transcontinental railroad. The family name probably had originally been Bueser. My father, Halcot Everett Roberts, mainly Welsh, was born in 1883 in Rosalia, Washington. His parents had come west by covered wagon. Both families settled in Washington Territory, which became Washington State in 1889, and raised wheat. The steep hills required combined harvesters, drawn by teams of twenty-four to thirty-two horses, to mow and thresh the wheat, then leave it in trails of two-bushel bags.

Fig. 2.1. *The Boozer Ranch, near Rosalia, Washington, 1975. Courtesy of M. A. Bowman.*

Fig. 2.2. *Combined harvester drawn by thirty-two horses, Boozer Ranch, ca. 1915. Courtesy of M. A. Bowman.*

My dad worked on his parents' wheat farm as a boy, plowing, seeding, and harvesting wheat, loading wagons with two-bushel sacks of grain—each of which weighed about 120 pounds. But he decided not to become a farmer himself. About 1906 he entered the Washington State College School of Pharmacy. In 1908 he felt ready to get married.

FIG. 2.3. *Author's parents, Halcot E. and Eva B. Roberts, ca. 1915.*

FIG. 2.4. *Author at age four, Rosalia, Washington.*

He didn't have to look far for a wife; he and Eva Boozer had grown up together and gone to school in Rosalia, a town where everybody knew everybody. After graduating he purchased a drugstore in Rosalia. My sister Margaret was born in 1909, and I came along in 1911, followed by Louise in 1914 and Fred in 1916.

Dad wasn't really a scientist, but he would have made a good one. He discussed with me, my brother, and sisters every new scientific discovery that he read about in the newspaper. While we might not have understood all of them, such things broadened our horizons. I remember vividly Dad's explanation of the "Heaviside layer" and how it moved upward at night, extending the range of radio stations, and down in the daytime, shortening their range. I had been born into the right family at the right time.

Mother went to Whitman College in Walla Walla, Washington, where she studied music. During the summer wheat harvest she cooked for the harvest crew, and the reputation of her pies and cakes was known for miles around. Besides being an excellent cook, Mother was also a great practical joker. One April first, she mixed cotton into the pancake batter. We kids chewed and chewed—until we guessed the reason for the new pancake formula. Then we all had a good laugh.

We lived in Rosalia until 1920 when Dad's health began to fail. The long hours in the drugstore had begun to wear on him. So we moved to Wenatchee, where Dad went into business selling Edison phonographs. At this time the Edison was by all odds the best, and Dad did quite well until the advent of radio in the early 1920s sent the phonograph business into a tailspin. Dad fought the growing popularity of radio for four years, but when his shop inventory was destroyed in a fire, he began to look for a new vocation. He didn't want to go back to being a pharmacist.

When Dad traveled to the towns in the Okanogan Valley to sell phonographs, he would sometimes take me along. One time when our fortunes were especially low, I accompanied him on one of his selling trips. While staying at Oroville, near the Canadian border, we wandered out onto a bridge over the Okanogan River to watch some men spearing salmon. Dad noticed that one spear was not being used, and he asked if he could borrow it. The owner agreed, and Dad tried his hand at spearfishing. Within a short while, he had speared a three-pound salmon. The spear had no barbs so the salmon jerked its way to freedom, but I followed it into the shallow river and retrieved it for our dinner. The salmon proved to be mighty good.

On another sales expedition, we passed through Omak in eastern Washington, and Dad saw a FOR SALE sign in a candy store and soda fountain. Dad walked in and surveyed the place. Evidently, he liked the layout and the equipment because he promptly looked up the owner and offered him six thousand dollars for the building and the business. The owner, Darrel Pinkney, looked at Dad and nodded; the two men shook hands, and the Roberts family suddenly owned a candy store and soda fountain. And so it was that we moved to Omak in the summer of 1924. Margaret and I went ahead with Dad to the new store, and Mother, Louise, and Fred joined us later after selling our house in Wenatchee.

We started out selling Dad's incomparable peanut brittle, nougats, pinoche, divinity, caramels, butterscotch, taffy, and fudge and drinks from the soda fountain. Gradually Dad responded to customers' needs and expanded the store into first a sandwich shop and eventually a restaurant. Mother, with her experience cooking for harvesting crews, became our first cook. The Peacock Cafe's reputation quickly spread north and south along the Okanogan River. All the family worked in the business. I would finish high school classes and go straight to the café to put in a long shift each day and even longer ones on weekends. I reigned over the soda fountain as official ice cream maker and principal soda jerk. I developed my own specialties—concoctions of banana splits, parfaits, ice cream sodas, and malted milks. Girls would come from miles around to see me perform. I loved being on stage, but once off, I lost my confidence—I was afraid of the girls!

A few times I did get up the courage to ask a girl for a date, but I seldom had the resolve to ask her out again. There was one girl I couldn't get out of my thoughts. I fantasized about delivering smooth and polished invitations, which I envisioned her eagerly accepting, but that was as far as I got with Marjorie Courtright. Marjorie (or Mij for short) and I were in school together. I had heard that she had a boyfriend at

Whitman College, but even if she hadn't, I wouldn't have been brave enough to ask her for a date.

Besides, all the long hours working in the family business didn't leave much time for extracurricular activities, although I did sing in a male octet—that is until Irene Kulzer, the music teacher, asked me to quit. What shame for a Welshman (even someone only part Welsh) to be unable to carry a tune. I felt that I had not lived up to my Welsh heritage.

During those early years, 1924–1928, while the business was getting established, Dad did his best to keep us from getting discouraged. He was full of zany ideas. Once he proclaimed he was going to train a clam to swim through the chowder to provide recycled flavor and lure customers to see the show. Another time he designed a rubber stamp to imprint a picture of a slice of ham onto a piece of bread. *You want a ham sandwich? Here's the bread, just stamp—and you have ham!* All joking aside, our food costs hovered at around 40 percent of sales, about double the average these days. No wonder we didn't make much money!

During the winter months, Dad and Mother went south to California to look for other work, leaving Margaret and me to run the café. While the responsibility was no doubt good for us, we learned early on how stressful it could be to rush the daily receipts to the bank just in time to cover the checks we had written!

While learning these important business skills, I tried not to neglect my studies. Two fine teachers, Robert Clemens and Charles Cooley, helped keep me motivated. Clemens's excellent teaching of chemistry gave me a strong foundation for university-level chemistry classes. Cooley taught general science. He liked working outdoors and arranged field trips for some of us who were willing to help him collect rocks. He had found a deposit of thulite, a bright pink mineral of the epidote family, on the Colville Indian Reservation. It is a finely crystalline mineral (hydrous calcium-manganese silicate) that forms in lenses a few feet thick in bedrock. It is not hard enough to be used in jewelry, but large slabs make attractive decorative stone. He enlisted the aid of several energetic boys in mining the thulite, which

Fig. 2.5. *Author upon graduation from Omak High School, 1928.*

he used as facing for a fireplace. (The thulite, streaked with green epidote, was beautiful, but the overall effect looked very pink.) Cooley awakened my interest in geology. I have always been grateful to these two teachers. They were there just when I needed a nudge to find my heart's interest for a career.

There were others who influenced my early interest in rocks. "Dad" Hayes, a handyman who worked in our store during the late 1920s, used to take me to the hills on free days to look for placer gold. Hayes had worked placer gravels in Alaska in 1898 and made a small fortune, but he had lost it prospecting for a second fortune. Hayes told me stories of his Alaskan experiences, the details of which I have long since forgotten, but he painted a rosy picture of gold as a source of instant wealth. Both sides of my family dabbled in mining. One of Mother's uncles, Charles Hurlburt, operated a silver mine in Ouray, Colorado, and several uncles and cousins on my father's side were involved in mining ventures. I heard many accounts of these operations; thus, I was conditioned for a career in mineral exploration from an early age.

I graduated from high school in 1928 without distinction. That fall, lack of funds forced me to delay entering college. I worked as many jobs as possible in addition to helping manage the café. I harvested apples, receiving five cents for every fifty pounds. I could earn about five dollars a day, which was good money in those days. After the picking season, I worked in warehouses, where the going rate was only about twenty cents an hour. There I moved boxes of apples from delivery trucks or loaded apples into freight cars to be sent to a cider plant. By living at home I saved enough money to go to Washington State College the next year.

University Years

My dream had always been to attend the University of Washington, Seattle (UW), but I was rejected because of my low grade point average. At Washington State, I enjoyed my chemistry courses the most until an elective course in engineering geology changed my life forever—I knew immediately that geology was to be my field.

I finally transferred to the University of Washington as a junior in 1933. Early on, I reconnected with two of my high school friends from Omak, brothers Quin and Stoddard De Marsh, already in their second and third years, respectively. They squeezed me into their apartment on the northwest corner of campus. The three of us paid our rent by working as elevator operators during rush hours, and we took restaurant jobs for the free meals. I found other part-time jobs so that I needed only a small amount of money from home, about 150 dollars a quarter. I generally took a full academic course load as well, and I learned to get by on little sleep.

While I was far from the best student at the university, I learned that if I studied hard and used my wits, I could compete with the best. For instance, in scientific German class, I noted that the professor jumped ahead to the next chapter when choosing passages to be translated in class and on quizzes, so I always stayed a couple of chapters ahead of the class. My game was almost my undoing, however, when it came time to take the German examination for the master's degree. I was given a passage in an unfamiliar book and did not do nearly as well as usual, but I received a passing grade nonetheless. I never did inform the professor of the reason for my sorry performance.

I used another gambit to win in chemistry. I had realized early on that I did not want to spend the rest of my working life in a laboratory, so I had made geology my major but continued chemistry as my minor subject because I thought it could be useful to a geologist. In a course in volumetric quantitative analysis, I completed my laboratory work for the quarter in six weeks and received all A's. As the top student, I was excused from taking the final examination; I was free to spend more time studying for my other final exams. I discovered that it's easy to beat the system if one can only find the right formula and has the resolution to succeed.

I found myself becoming ever more absorbed in geology, and the reason had to do, more than anything else, with the inspiration I received from UW's notable, well-balanced geology faculty members: Professor George E. Goodspeed, the head of the department, who taught petrology (the study of rocks); Howard Coombs, who was strong in volcanic geology and the geology of Washington State; J. Hoover Mackin, who was a powerhouse in geomorphology and regional geology; Charles E. Weaver, who had wide experience in western Mesozoic and Tertiary stratigraphy and paleontology; and Julian D. Barksdale, who excelled in classical geology and stratigraphy of the United States. UW's geology students received a strong foundation in the basics of geology. These five men launched my career.

Mackin came to UW during my senior year. He immediately brought another dimension to my studies, adding geomorphology, the study of land forms, to stratigraphy and structure. His lectures brought geology to life for me. Mackin often recited a favorite quotation from Lewis Carroll's *Through the Looking Glass,* challenging us to think and to use our imaginations.

I can hear him now: "I can't believe that!" said Alice. "Can't you?" the [White] Queen said in a pitying tone. "Try again; draw a long breath and shut your eyes." Alice laughed. "There's no use trying," she said: "one can't believe impossible things." "I dare say you haven't had much practice." said the Queen. "When I was your age, I always did it for half-an-hour a day. Why, sometimes I've believed as many as six impossible things before breakfast."

The dialogue was catchy, but I couldn't immediately see the point. How could one believe six impossible things at a time, especially before breakfast? I would come to understand that Mackin really meant we should keep our minds (and imaginations) open and consider all the alternatives when solving problems. James Gilluly, in his memorial to Mackin, said, "He insisted always that his students reason out and understand geologic ideas, not merely parrot them." This proved to be a prophetic statement—one that would come to my mind many times in the future. Students of J. Hoover Mackin were lucky to have been part of his world.

I was a teaching assistant during my last year at UW and finally graduated with a bachelor of science degree, cum laude, at the end of the winter quarter in January 1935. Perseverance paid off!

In addition to being wonderful teachers, Coombs and Barksdale offered me crucial help in advancing my career. Coombs gave me a boost into the U.S. Geological Survey. In 1935 he recommended me for a summer job with J. B. Mertie Jr. in Alaska. In those days jobs like that were almost impossible to get—I had tried without success. Few people today can comprehend how great a gift this was! Barksdale later recommended me for the doctoral program at Yale, which propelled me into the stratosphere.

As soon as I learned that Coomb's recommendation had landed me a job offer with the USGS, I called Mertie to accept, and shortly thereafter, in summer 1935, I was on a steamer to Alaska to work as a geologic assistant northwest of the Alaska Range. In Anchorage I met the rest of the project's personnel: Mertie, a boatman, and a cook. We flew from Anchorage to our field camp. Our flight path took us over the rugged Alaska Range through Rainy Pass and across Lake Iliamna to the Tikchik Lakes. We carried mostly dried food, supplemented by a good deal of Gruyère cheese (J. B.'s favorite) and milk chocolate. The plane landed smoothly on Lake Chauekuktuli, and we set up camp on the shore.

We began fieldwork there and then gradually made our way south to Lake Nushagak. The regional geology was dominated by greenish gray sandstone, called graywacke, interlayered with gray shale, cut by an occasional small granitic intrusive. It was a fairly monotonous sequence of rocks, and we mostly plotted the strike and dip of the beds in our attempt to work out the major folds in the area.

It was a great summer close to the Arctic Circle, where the sun shone most of the night during June and July. Although the Tikchik Lakes are in Alaskan brown bear country, we never saw one. We frequently saw bear signs and tracks along streams as well as partially eaten carcasses of salmon—it was a sockeye salmon spawning area. I was constantly on the alert, and at first I carried a rifle on traverses into the hills for protection. This was fine until the day I got into a tight place on a granitic ledge and had to jump to a lower ledge, nicking the rifle barrel in the process. That was when I put the rifle into safekeeping.

In the Tikchik Lakes region, Mertie would retire to his tent on rainy days and work calculus problems to pass the time! Because Mertie weighed about 240 pounds, he confined his geologic mapping to areas easily reachable by boat while I climbed steep slopes and ridges. It was a good division of labor; at twenty-four years I had energy to burn, though climbing up over alders flattened by snow was no fun.

We supplemented the dried food that we had brought along with fish that we caught, and we kept a lookout for other kinds of meat. One day I saw a small band of woodland caribou. I reported this to our boatman, who killed one, and for a while our diet was enlivened with excellent caribou meat.

One incident of that summer remains a vivid memory. I took off alone to Lake Chikuminuk, about seventeen miles by trail from Lake Chauekuktuli. The trip went well until I had to wade through a fast, icy stream. Boulders in the streambed made the footing precarious, and I had a hard time keeping my balance in the knee-deep water. If I had slipped, I would have become thoroughly chilled. Fortunately, I made it across without a problem. I reached my objective, made notes about the rocks along the lake's edge, and headed back. It was growing late, and although I retraced my earlier path almost exactly, I arrived at the meeting place a couple of hours late. I assumed that the boat had come for me, and when I wasn't there, it had returned to camp. I sat down on the bank of a small pond, and for the next ten hours, I watched a beaver swimming back and forth, occasionally slapping the water with his tail, and a loon paddling in broad circles in front of me and raising its head to give its eerie cry. About 7:00 A.M., Mertie, the boatman, and the cook showed up with a stretcher cut from alder boughs, certain that I had run into trouble. I was a much chastened lad who rose and said, "I'm sorry, I guess I bit off more than I could chew."

The summer's climax was a train trip from Anchorage to the village of Tanana on the Tanana River, then down the river by steamboat to the Yukon River and up the Yukon to Dawson and on to Whitehorse, where we caught the train to Skagway, thence to Seattle by steamship. What a trip for a budding geologist!

In 1935 a number of undergraduates at UW took the civil service examination for junior geologist. As I recall, four of us passed: Warren Hobbs, Charles Carlston, Allen Carey, and me. Since many questions required a knowledge of petrography and petrology, UW students had a clear advantage. We were successful largely because our undergraduate curriculum included petrography (microscopic study of thin sections of rocks); most universities offered it only as a graduate course.

As a junior geologist, I was offered a position in summer 1936 assisting Stephen R. Capps in the Dixie district of Idaho. The Dixie district was a gold mining area, containing both placer and lode mines. Mostly I helped make a topographic map of the active placer area, but near the end of the season Capps permitted me to map some of the lode mines. This was not a major project, but it was my own, and it dealt with gold. Capps wrote a pamphlet for the Idaho Bureau of Mines (1938) on the district and included a short section on the lode deposits, and while it was minor, it was my first published report.

At the end of the Idaho field season, I returned to UW to complete a master's thesis on the summer's work and to take courses in engineering. I thought it would be useful to know something about mining equipment, assaying, mineral separation, and flotation processes, as I could see jobs in geology were mainly in the mineral industry.

In the late 1930s, jobs in geology were rare, so some students considered advanced graduate work. Although I had received offers of scholarships from several universities, including Harvard and Stanford, I was strongly influenced by Barksdale, whose own Ph.D. was from Yale University. At that time my principal interest was in petrography, and as Adolph Knopf at Yale was the best in that field, Barksdale offered to help me and a colleague, Warren Hobbs, go there. We both needed financial aid, and Barksdale was willing to assist us in securing teaching fellowships. Barksdale felt, however, that he could send only one at a time. So, it was decided that Hobbs should go first and that I should remain at UW and write a master's thesis.

In the spring 1937 I was asked by the U.S. Geological Survey to go to Arizona to assist Sam Lasky in mapping the sedimentary manganese deposits near Artillery Peak on the Bill Williams River, a tributary of the Colorado River. It is Mohave Desert country, low in altitude and hotter than Hades in late spring and summer. The inhabitants are mainly creepy, crawly things like rattlesnakes, lizards, scorpions, Gila monsters, and tortoises; I respected them but mostly ignored them, although I always jumped when I heard the telltale buzz of a rattler. The temperature was tolerable in March, and we got along with the mapping project fairly well; I again was a topographer. As the spring wore on, and the heat built up, reading a stadia rod over distances of more than a thousand feet became almost impossible due to the dancing heat waves—at least the numbers on the stadia rod danced. As you would expect, certain inaccuracies crept into our records. Although I tried hard, at times Sam was not real happy with me, and I didn't blame him. I would have given a lot for a global positioning system back then.

Along toward the early part of May the temperature began to soar. When it reached 121 degrees and stayed there overnight, we could no longer carry enough water to last through the day. So, with most of the work complete, we called the job done.

I was then asked to drive a car from Phoenix to Seattle for use in a manganese project under the direction of Charles Park in the Olympic Mountains in Washington State. There I would work with John Nelson, formerly a University of Washington student. (John Robinson, also from the university, continued our work after Nelson and I went to Nevada in 1939.) The Olympic Mountains are almost the exact opposite of Arizona; they are densely forested and receive from sixty to three hundred inches of rain annually. I assisted in geologic mapping while looking for manganese deposits, starting each day near sea level, climbing to five or six thousand feet during the day, and descending to sea level along a different route in the late afternoon. My hikes were enlivened by glimpses of deer, elk, and bear. One day I was walking through a patch of huckleberry bushes and came face-to-face with a black bear that had been picking huckleberries. I turned slowly and started walking down the slope; the bear turned slowly and started walking up the slope. After a few steps, I started walking faster, and when I glanced back, I saw that the bear had done likewise. Neither of us wanted to tangle with the other. I didn't want to test my geologic pick against his claws.

The Olympics gave me my first look at sediments that had been deposited in deep water along with interlayered submarine volcanic rocks. Little did I know that, in another place at another time, oceanic rocks deposited in a similar geologic environment would dominate my life as I searched for ore deposits in Nevada.

But my immediate future was at Yale University. Warren Hobbs and I boarded a train in Seattle for the trip to New Haven, Connecticut; it was my first trip outside the West. We arrived the day after the 1938 hurricane, which had devastated much of coastal New England. People were beginning the backbreaking cleanup: repairing the white clapboard homes and removing the venerable oaks that had withstood many other storms before falling to this one. New Haven had an eerie feel on this clear, sunny morning.

Once I was enrolled, Adolph Knopf, director of graduate studies, arranged a curriculum for me that emphasized petrography, advanced general geology, structural geology, and economic geology. I particularly wanted to study petrography, and Knopf wisely pointed out that most jobs for geologists in those days were in economic geology, though petrography was very useful in economic geology.

Fortunately, at Yale I received full credit for my previous year's work at UW. I was asked to spend only one year in residence plus a year for my dissertation. And since I had received a passing grade on the civil service exam, Knopf felt I should not remain in school any longer than absolutely necessary. Knopf outlined my duties as teaching fellow, for which I received tuition and a generous six hundred dollars a year.

One of the critical times in a graduate student's life is the oral examination—the orals. At Yale, orals were usually given at the end of the second year of graduate school, but as I had received graduate credit for my work at UW, I took the orals near the end of my first year. The orals are a make-or-break hurdle for graduate students. If a student passes, he goes on to complete a dissertation for a Ph.D. If he fails, he can settle for a master's degree or change his course of study. I was extremely relieved to be able to answer all questions leveled by five professors, even a difficult crystallographic one asked by the dean, Charles Warren, who loved to fluster students. I have long since forgotten the details of his question, but I remember that he gave me some crystallographic parameters and asked me to reconstruct the crystal and tell him the name of the mineral. I was able to describe a crystal that fit the dean's parameters satisfactorily. Also, I speculated that the crystal could be a mineral of the pyroxene group (a calcium, magnesium, and iron silicate), a mineral that might be found in certain lavas. This answer was accepted by the dean. Having surmounted this challenge, I passed the orals with honors.

Adolph Knopf taught advanced general geology, a course designed to take students back to times when little was known about the world and to discuss fundamental problems such as the age of the earth and the evolution of ideas in geology. His approach encouraged students to expand their horizons and to conceptualize the big picture by reaching beyond the sometimes limiting views of the day. Knopf was an expert in petrography and petrology, and I now know that it was a rare privilege to be a student in his classroom. I will never forget the awe I felt when he first described an exploding atom; the theory we discussed that day would soon be tested in the great war against Japan.

At Yale I made many enduring friendships. In addition to Warren Hobbs, I found particular rapport with Preston Cloud, John Rodgers, and Hugh Beach. We decided that living together would keep expenses low, so we rented an apartment over Menzies Plumbing Shop. Rodgers later joined the Yale faculty as a professor of structural geology.

Another great friend was A. Lincoln Washburn and his wife, Tahoe. Washburn had set his sights on a professorship in geology, and he finally achieved it at the University of Washington, where he served with distinction.

Another member of the Yale faculty was Alan M. Bateman, who taught economic geology. I was to work with him on the Board of Economic Warfare during World War II. Alan became my model for professionalism and confidence. He often told this story: "The world's three greatest economic geologists were standing on a hill in South Africa. I said to the other two, Waldemar Lindgren and Louis Graton, 'This is going to be a mine.'" Bateman exuded confidence, and he taught all of us to believe in ourselves. He drilled into us the notion that geology told the simple story of the evolution of the geologic framework of a region. Chester Longwell, my major professor, visited me in the field and enthusiastically supported my definition of the Antler Orogeny. Shortly after that, he retired, and John Rodgers competently assumed Chester's position on the faculty. John became my major professor and approved my dissertation.

The year sped by, and in June 1939, the academic work and lifestyle I had grown to love came to an end. I would never again enjoy such an idyllic period. I would happily have stayed in the safe cocoon of intellectual pursuit, but reality was harsh. Talk of war was beginning to occupy everyone's mind. I vaguely remembered the stories of the First World War, and it was difficult to imagine that a second could be coming. It had seemed so far removed from my small world, but now it could touch us all, and there would be no way of stopping it. But at least for now, I knew I had a job waiting with the USGS—for how long was anybody's guess. Warren Hobbs and I left Yale and headed home, knowing we might not be home for long.

Chapter 3

Sonoma Range

1939–1942

> What do I feel? The way you feel when you come to the top of a ridge and look down across miles and miles of land you have never seen (before). . . . the pleasant sound of running water, the way the leaves turn red in the fall.
>
> —LOUIS L'AMOUR, *Westward the Tide*

> He said there was one only good, knowledge; and one only evil, namely, ignorance.
>
> —DIOGENES

While I was working in the field, cresting the ridge was always important to me, for then I could take a breather, and look ahead across the valley. It generally happened about midday; then I would have lunch and plan the afternoon.

When Warren Hobbs and I left New Haven in June 1939, we took the train to Flint, Michigan, where we picked up a new car that we were to drive out to Mackin at the University of Washington. From Flint we headed for Helena, Montana; Warren wanted to visit his fiancée, Arleda Allen. They had met when Warren was working as field assistant to Maxwell Knechtel of the USGS in the Little Rocky Mountains and had become engaged the previous summer. When I met Arleda I was immediately impressed with her beauty, self-possession, and cordiality. Our eyes met and held for a long moment. She was wearing Warren's ring, but her greeting to him seemed restrained. She introduced us to her roommate—the girl I was supposed to escort around Helena for the next few days. I do not remember much about that visit, except that I was drawn to Arleda. Since she was Warren's girl, I tried not to let my feelings show, but I suspect that she knew I found her attractive, and I sensed that she was interested in me.

As Warren and I both had summer jobs with the USGS, we stayed in Helena only three days. When we reached Seattle, Mackin was a little puzzled about the excess mileage acquired on our trip, but Warren gave him a satisfactory answer. I only knew that I was enormously glad to have met Arleda.

A couple of weeks later I boarded a bus for Nevada, where I had accepted a position as junior geologist in the Metals Branch of the USGS under D. Foster Hewett. My first assignment was a new project in Nevada under the overall direction of Henry G. (Fergie) Ferguson. Our job was to map the geology of the Sonoma Range one-degree quadrangle (a quadrangle is a rectangular map area; a one-degree quadrangle runs one-degree latitude and one-degree longitude. At the latitude of Nevada, it covers about seventy miles from north to south and fifty miles from east to west, about thirty-five hundred square miles). We were to make the map, assemble a geologic framework of the region, and study the mineral deposits. I now had a full-time job with the USGS at a salary of two thousand dollars per year plus field expenses. Not bad for those days!

When I stepped off the bus on July 4, 1939, in Winnemucca, Nevada, a blast of hot, dry air and Henry Ferguson greeted me. He quietly said, "Hello, I'm Fergie." He was short, about five feet, four inches, and his unassuming, almost shy personality immediately put me at ease. I soon learned that his self-effacing manner hid a great intellect. Siemon (Si) Muller, professor of geology (paleontology) at Stanford University, was with Fergie. I sensed that I was in the right place to learn a lot of field geology.

We went to Bob's Cafe in Winnemucca for a quick lunch, and Fergie exclaimed, "We want to go out to the East Range to look at the Triassic rocks there." Si, a specialist in Triassic rocks and faunas, had worked in the Sonoma Range south of Winnemucca previously, but

MAP 3.1. *Principal mining districts and cities of Nevada.*

FIG. 3.1. *Author, 1939, Sonoma Range.*

he had never had a chance to look at the East Range. On that first, brief trip, we found what Si was looking for—Triassic formations identical to those he had studied in the Sonoma Range. I got a serious blister on my heel during the hike but didn't complain. I found out right away that we were going to tackle the geology of the Sonoma Range quadrangle head on! The rocks and structures we eventually were to map would change the face of Nevada geology forever.

Initially, my fieldwork responsibilities were to study and map the igneous rocks, both volcanic and intrusive. I also studied mineral deposits, principally gold- and silver-bearing veins. When time permitted I helped Fergie map the Paleozoic rocks that proved to be critical elements of the regional geologic framework. Map 3.2 shows the distribution of the rocks before the deformation that brought the oceanic rocks onto the continental margin.

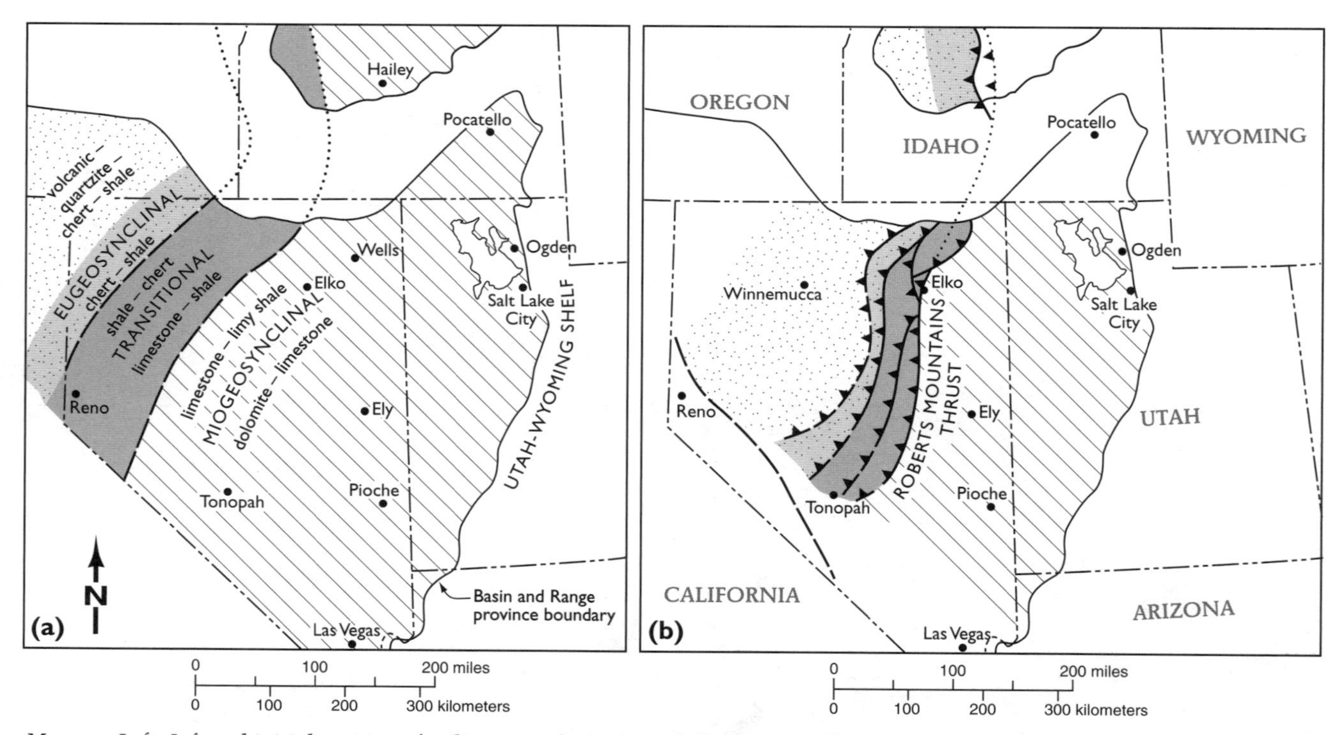

MAP 3.2. Left: *Inferred initial position of sedimentary facies in cordilleran geosyncline;* right: *distribution of facies after late Paleozoic thrusting.*

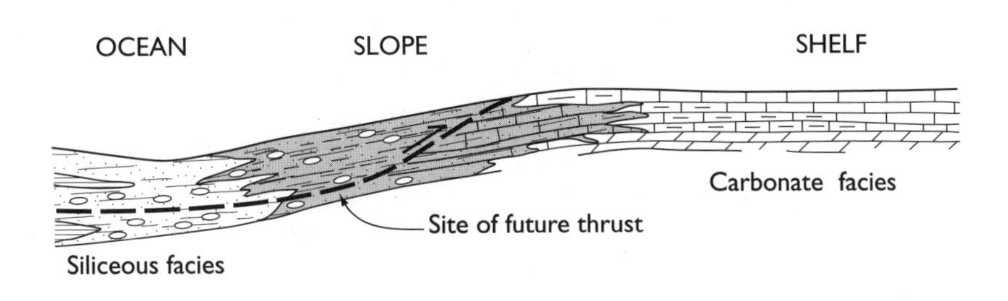

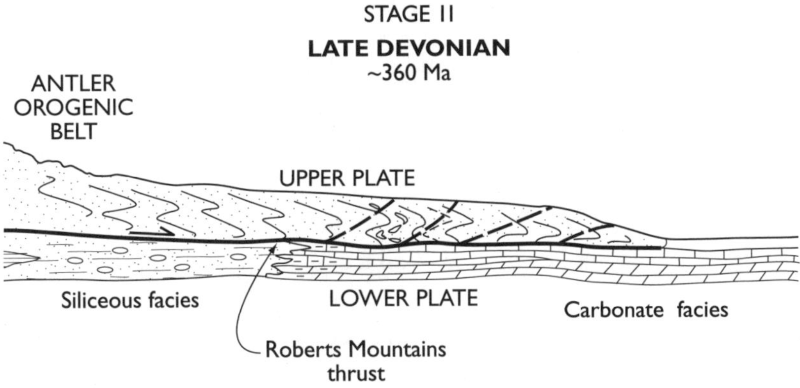

FIG. 3.2. *Principal stages in the development of the cordilleran geosyncline. Stage I. Deposition from Cambrian through Devonian time. Stage II. Distribution of facies after thrusting on Roberts Mountains thrust.*

Henry G. Ferguson

Fergie was first and foremost a field geologist. For a time he had been in charge of the USGS Metals Branch in Washington, D.C. Fergie worked primarily in the Great Basin and authored many important papers on its mining districts. He was largely responsible for setting up the geologic framework of Nevada as we know it today.

My association with Fergie permitted me to see how his agile mind worked and how he solved complex structural problems in the field. Initially, I wondered about the relevance of some of his far-reaching predictions, but looking back I see that from him I learned how mineral deposits fit into the framework. Fergie insisted that the best way to solve a geologic problem was "to map it," and the solution would miraculously appear.

In the Field—Pay Dirt

The first inkling that the Sonoma Range quadrangle might hold the key to the regional geology came in Willow Creek in the East Range. Entering Willow Creek from the west, we first encountered the Inskip Formation, which contained a coral of Mississippian age (about 330 million years ago) and which rested unconformably on the Leach Formation, a chert, quartzite, shale, and volcanic unit. The Leach had been thrust over a unit of limestone and shale. Initially, Muller thought that the limestone and shale might be Triassic in age, thereby dating the thrust as post-Triassic. But later work by Whitebread (1987) revealed that they were early Paleozoic in age, thus permitting us to date the thrust as pre-Inskip and therefore Mississippian or older. This proved to be a critical element in the regional geology.

In Willow Creek we visited Wallace Calder, a dentist from Winnemucca who had mined very handsome nuggets, some weighing as much as twenty ounces, from a placer deposit. Calder said that these nuggets came from a fault zone in Willow Creek and showed us the fault in the Wadley mine that he owned. We later determined that it was a thrust fault that brought oceanic-facies rocks (rocks from deep water) over limestone of the continental shelf. We did not know whether these events were merely of local significance or whether they signaled a larger regional pattern, but we recognized two important facts: (1) deep-water oceanic rocks had been thrust over a continental margin in shallow water, and (2) gold occurred in and below the thrust zone. These were only the first steps in assembling a geologic framework in which gold was to be found. Following Mackin's advice, taken from *Through the Looking Glass,* we were believing in impossible things before breakfast. We now had two of them.

When in the field, Fergie and I generally worked the same area. It didn't take me long to become totally hooked on the rocks of the region. Certainly they were new to me, but they gave tantalizing

FIG. 3.3. *Gold nuggets from Calder placer, Willow Creek, East Range, about forty-five miles southwest of Winnemucca. The dark gangue mineral is quartz. Courtesy of Lois Calder Baum.*

glimpses of spectacular formations folded into complex structural forms. They reminded me of photographs and drawings of the Alps I had seen in textbooks. Like Alice in Wonderland, my curiosity was aroused, and before I knew it, I had fallen down the rabbit hole, never once considering how in the world I was to get out again.

Strategic Minerals — D. Foster Hewett

Later in 1939 funding came from Congress to study certain strategic minerals that the United States lacked or produced only on a small scale, including mercury, antimony, tungsten, nickel, manganese, chromite, beryllium, and vanadium. This program was instituted by D. Foster Hewett, chief of the USGS Metals Branch. Foster's interests included mineral economics. Several years before he had foreseen shortages in strategic minerals and had formulated a plan to study the occurrences of those elements, as well as such commodities as mica and quartz, in the Western Hemisphere. He had outlined his projections in a series of lectures at the Brookings Institution in Washington, D.C. The support he received there smoothed the way for Congress to pass a funding proposal for evaluation of deposits of these materials. He had started several projects to evaluate the deficiencies, and I had previously worked on two dealing with manganese—one in Arizona with Sam Lasky and another in Washington's Olympic National Park. I was one of the first geologists to be chosen for the program. (Others who joined the Geological Survey at this time included Ogden Tweto, Warren Hobbs, Paul Averit, Lincoln Page, Donald White, Robert Bryson, Philip Guild, and Richard Jahns.)

Now, in addition to my work on the Sonoma Range, I was asked to map, evaluate, and report on deposits in the following areas: quicksilver (mercury) at Bottle Creek (1940) and Buckskin Peak (1940), manganese in the Nevada district (1942), and tungsten at the Rose Creek Mine (1943). These deposits proved to be small, but it was necessary to study and evaluate them for possible future emergencies. My reports were short and to the point, but writing them helped prepare me for later, more detailed work with the USGS.

Sonoma Range Quadrangle

An early chore related to our Sonoma Range fieldwork was writing sections of a report on the economic geology and intrusive (such as granite) and volcanic (basalt and andesite) rocks of the quadrangle. Fergie focused on the Paleozoic details, Si, on the Mesozoic. They collaborated on the structural geology section, recognizing that the Paleozoic and Mesozoic rocks had been folded and thrust faulted in both Paleozoic and Mesozoic times. I was keenly interested in their work; many times, when there were no ore deposits near one of our camps, I would help Fergie with his mapping of Paleozoic, Mesozoic, and Cenozoic volcanic rocks.

As we worked on the reports, it became increasingly obvious that a 1:250,000 scale (one inch equaling four miles) would not be adequate to show the detailed geology we had mapped. So, Fergie decided to split the Sonoma quadrangle into four parts, which he would present in four maps, each using 1:125,000 scale. These were named the Winnemucca, Golconda, Mt. Moses, and Mt. Tobin quadrangles. Authorship was assigned on the basis of the principal contribution of each person. The Antler Peak quadrangle, called the fifteen-minute quadrangle because it used a scale of 1:62,500, was separately mapped because it contained the major ore deposits in the Sonoma Range quadrangle. This was my responsibility, and I later used the project for my doctoral dissertation at Yale.

The mapping of the Paleozoic rocks led to the most important project of my early career—the definition of the Paleozoic facies that, by magical chance, just happened to be best developed in north-central Nevada.

Through Fergie's eyes and mind, I was treated to a detailed analysis of these features. For example, Fergie took each rock exposure and broke it down into its component bits and pieces for me, explaining what each meant in the regional story. Fergie would determine the direction my geologic career would take, and there couldn't have been a better instructor. During our early work in the Sonoma Range we saw highly deformed rocks, locally overlain by undeformed (not folded) conglomerate, sandstone, and limestone, but these exposures were small and disconnected, and we could not assemble a coherent picture.

We were seeing only two of the many critical geologic features in the Sonoma Range, which included highly deformed Paleozoic rocks of oceanic origin, complexly folded and faulted. These rocks also contained some minor gold deposits along a thrust fault. We could see all these features so clearly, yet they did not add up to a complete picture. We continued to hope that a clue to the whole would reveal itself. Sometimes the harder we looked, the more confusing the puzzle became. Yet Fergie, like the White Queen, was always encouraging Si and me to keep looking, to keep imagining the impossible.

Washington, D.C.—1939

In the fall of 1939, we completed our fieldwork for the season, and Fergie and I left for Washington, D.C., where we would spend the winter writing our reports. Si Muller returned to Stanford. In Washington the New Deal agencies were being closed and replaced by agencies that would coordinate the war effort if we joined the fighting in Europe. I sensed that everyone was on edge. British people I met seemed certain that the European fracas had the potential to be World War II and that the United States would be involved.

Despite the charged atmosphere, or maybe because of it, those were great days to live in Washington. The grim possibility of war was a stark reminder that one had better enjoy life today. I soaked up as much city life as I could, going to many good and inexpensive restaurants scattered throughout the city. There was excellent seafood in places along on the waterfront, and throughout the district were superb ethnic restaurants—Indian, South American, French. There was no shortage of places to visit—historic sites, museums, the National Symphony, the Smithsonian, Capitol Hill, and many others. At times a company from the Metropolitan Opera came to Constitution Hall, which was quite inadequate for such performances, but the beautiful voices more than made up for the poor staging.

I was completely on my own with the princely salary of two thousand dollars a year, not a lot of discretionary income, but those were the days when a fairly good suit could be bought for twenty-five dollars. So all things were in proportion. Often I shared a room in a boardinghouse near Nineteenth and G Streets with another junior geologist. The rooms cost only about thirty dollars a month, and it was easy to go back to the office evenings and weekends to put in a few more hours on reports.

My social life was minimal when I first arrived, but I soon surmised that there might be hope for improvement if I could learn to dance. I went to the Arthur Murray Studio, where I learned the fundamental fox-trot and waltz and also a good Cuban rumba—which requires keeping the weight on the back foot while advancing the other foot with a relaxed knee and results in a very successful gyration of the hips. Fortunately, Cuba was our friend at that time, so it was politically correct to dance the rumba. I also learned an Americanized version of the samba and the Argentine tango, but I never did master the polka. Often I would invite my instructress to go to one of the small dance club bars (located in the best hotels) where top bands played. These evenings were inexpensive; drinks were fifty cents to a dollar a shot, and inexpensive taxis carried us from place to place. As I gained confidence in both my dancing and my social skills, I felt safe enough to invite other girls to go the same route.

On weekends, if the weather was good, we would drive out to Virginia for hiking and picnicking along the Appalachian Trail or rent canoes and paddle along the Potomac River at Great Falls, which was especially exciting in the early spring when the water was high.

Although I was enjoying myself, I was aware that war clouds were steadily gathering. America had not yet joined the conflict, but we knew that we soon would be involved one way or another.

Chapter 4

The Antler Orogeny

Heaven is high, and the Emperor is far away.

—CHINESE PROVERB

The swiftest traveller is he that goes afoot.

—THOREAU, *Walden*

Battle Mountain

The Battle Mountain Range in the Antler Peak quadrangle gave us critical clues concerning the geologic framework of Nevada. It contained clear-cut evidence of a late Paleozoic orogeny that had played a major role in the distribution of the rocks and control of gold deposits throughout the region. Battle Mountain was not one of the high ranges in this part of Nevada, but it literally had a heart of gold. It contained a major copper-gold deposit at Copper Canyon. Thus, analysis of the geology of the range yielded critical elements to our understanding of the geologic framework of north-central Nevada.

Antler Peak

Antler Peak stands proudly above the eastern ridges of Battle Mountain Range and is visible from Galena, the site of rich silver mines of the late 1800s. No one knows how Antler Peak got its name. I have wondered if it might have been named for the upward cant of the limestone that forms it or possibly for the many deer antlers found on its slopes. It has been called Antler Peak at least since the Fortieth Parallel Survey was published by King in 1878. His map showed Antler

Peak as Upper Coal Measures (Pennsylvanian) resting on Lower Paleozoic rocks.

Knowing that significant copper-gold deposits had been mined in the range, Fergie had requested that the topography of the Antler Peak fifteen-minute quadrangle be mapped at a scale of a mile to an inch so that the geology could be shown in more detail. When Fergie told me this, I immediately wondered why the ore deposits were more significant here than in other parts of the Sonoma Range. I asked him if he thought there was a geologic reason, or was it mere chance? Fergie told me right away that the element of chance did not enter into geology. "First," he said, "look for a reason, make a list of possible explanations, select likely answers. Then a logical pattern might develop. Geology is an orderly science, but you have to map the rocks before you can get the answer." That was lesson number one.

Here I was, fresh out of Yale, having completed the course work for my doctorate but needing a dissertation topic, preferably a field project. Could Battle Mountain and its ore deposits provide it? Although work on the western Sonoma Range occupied most of my time, I did steal a day here and there to scout the Battle Mountain Range, taking photographs of interesting areas.

Early in the summer of 1941 I planned a quick look at the Antler Peak quadrangle. I had asked Fergie's camp cook to drive me to the old silver mine at Galena. It was easily reachable via Duck Creek Canyon on the eastern side of the range, and I arranged for him to pick me up much later in the day at a ranch on the west side of the range. So, I took off on my first reconnaissance across the range. I first walked westward, on a course that would take me north of Antler Peak.

On this traverse I climbed for the first time the imposing red cliffs of conglomerate (ancient gravels), which contained angular boulders, mostly of limestone, sandstone, chert, and quartzite, some more than a foot in diameter. The conglomerate dipped gently westward, overlapping a unit of highly deformed sandstone and shale. I did not know the age of these rocks, but I did know that prior to the deposition of the conglomerate, intense folding of the underlying sandstone and shale had taken place. This folding meant that a significant orogeny had taken place in this region.

FIG. 4.1. *Imposing cliffs of Battle Formation, north of Galena, Battle Mountain Range.*

FIG. 4.2. *Angular limestone boulders more than a foot long, Battle Formation.*

The conglomerate unquestionably was derived from the folded rocks and had been deposited by streams capable of carrying large boulders. This certainly signified a source terrain of high relief, perhaps something like the rugged coast of California's Big Sur region. As I continued west, I saw the conglomerate grading upward into pebbly beds with finer grains, and I noted thin limestone interlayers, which

FIG. 4.3. *Basal Battle Formation composed largely of Harmony Formation cobbles, resting unconformably on tilted Harmony Formation about a half mile south of Galena.*

FIG. 4.4. *Antler Peak from east. Courtesy of USGS Professional Paper 459-A by Ralph J. Roberts.*

became thicker on the slopes of Antler Peak, forming a massive summit composed of steplike beds, resplendent in the morning sun.

In the 1878 survey, King had determined the limestone to be Pennsylvanian in age. This meant that the sea had lapped onto the formerly rugged Antler Mountains and reduced them to mere nubbins of low relief, which told me that I had found the critical exposures to work out the geologic framework of the region. The folded sandstone and shale signified a period of intense orogeny. The area had been uplifted, furnishing coarse debris to raging streams, which carried it to a fringing sea. The orogeny in this area had happened about three hundred million years ago—in the Pennsylvanian, or late Paleozoic, era. Erosion had continued until the sea covered the former mountains. The orogeny was prelimestone—i.e., earlier in the Pennsylvanian epoch than the limestone. I continued west and noticed that the conglomerate began to overlap a unit of chert, quartzite, shale, and greenstone (altered volcanic rock). This, like the sandstone and shale unit, had been intricately deformed. Both units were therefore older than the conglomerate. As I approached the western margin of the range, the conglomerate and limestone pinched out, and another chert and shale unit flanked the south side of the ridge I was traversing. I could see a few fragments of limestone on the ridge, but the massive unit that I had followed earlier was gone.

Now I had seen two more pieces of the regional puzzle. The first two pieces, obtained from Willow Creek, were (1) that deep-water oceanic rocks had been thrust over a shallow-water continental margin, and (2) that gold occurred in and below the thrust zone. Here at Antler Peak I had seen (3) the deposits of the conglomerate and limestone of Pennsylvanian age, and (4) the effects of still later Paleozoic orogeny. Wow! What a picture. Stop for a moment and visualize these magnificent, mind-boggling events. As in Willow Creek in the East Range, here highly deformed oceanic rocks of Lower Paleozoic age had been thrust into the region. These rocks were overlain by conglomerate that I later named the Battle Formation. The conglomerate graded upward into the Antler Peak Limestone, which had already been dated as Pennsylvanian. These facts formed the crux of my dissertation. I had discovered a major orogeny and had clearly dated it as late Paleozoic.

It isn't often that one can describe an orogeny as the centerpiece of his dissertation. The Acadian Orogeny in the Appalachians, the Laramide Orogeny in the Rockies, and the Alpine Orogeny had all long been recognized and dated by geologists; this new orogeny had

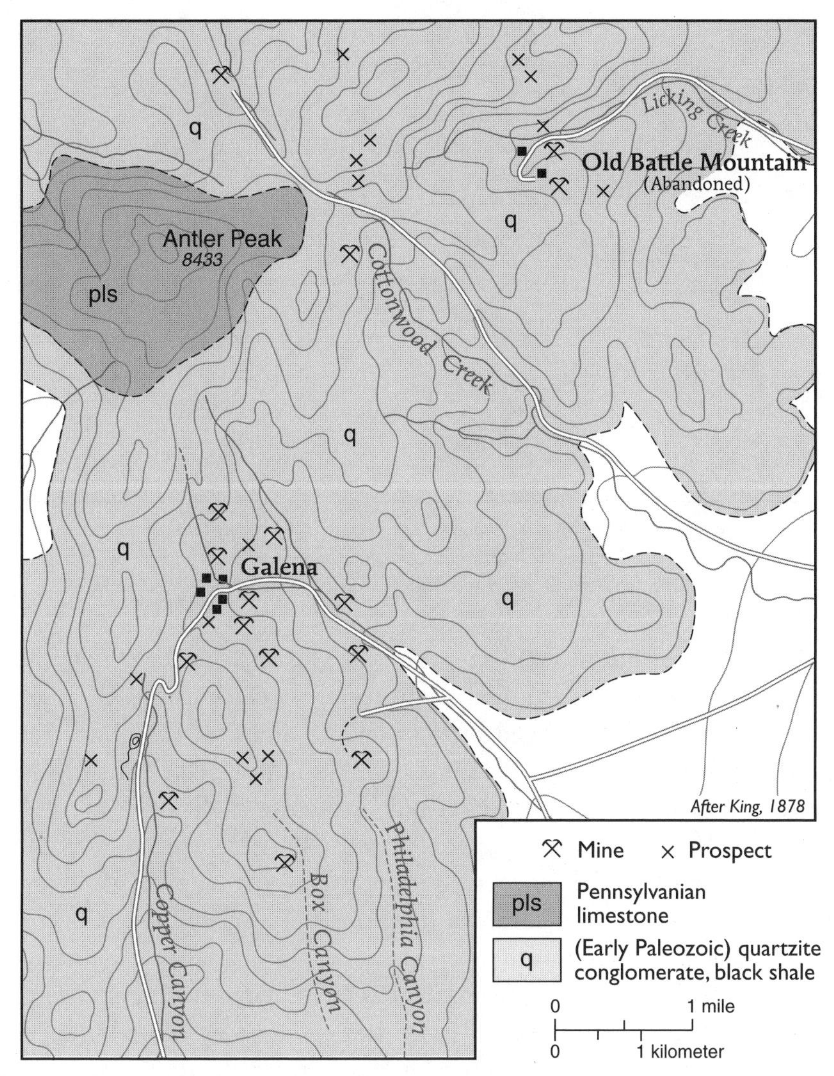

MAP 4.1. *This geologic map of Antler Peak area by Fortieth Parallel Survey (after King, 1878) shows Pennsylvanian Antler Peak Limestone, but grouping of underlying Battle Formation with older rocks of the Roberts Mountains allochthon fails to recognize critical evidence of the Antler Orogeny.*

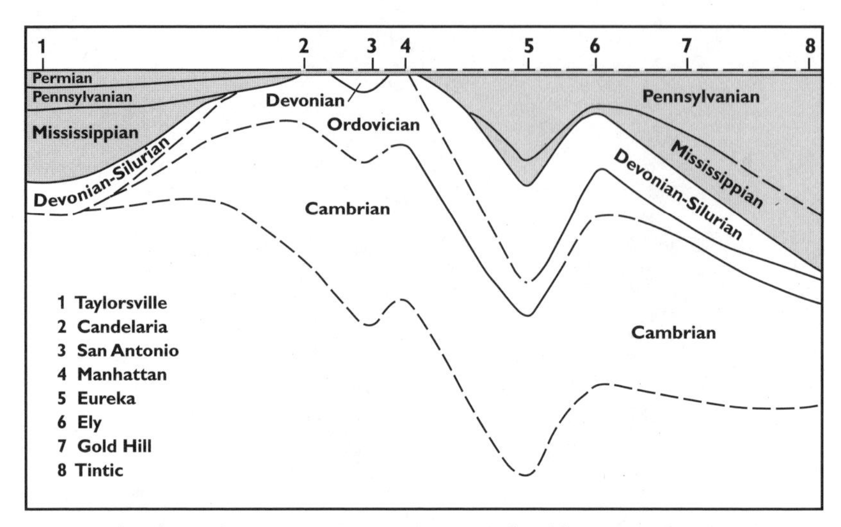

FIG. 4.5. *The "late Paleozoic positive area" as visualized by Tom Nolan (1928). Tom erroneously connected rock units across the "high" and did not recognize the predominantly oceanic rocks in western Nevada in early Paleozoic time.*

been suspected but never clearly defined. The rocks in the core of the range were well exposed, and some regional relationships that had been only poorly preserved in the Sonoma and East Ranges to the west were neatly arrayed here in Battle Mountain, just waiting to be interpreted and described. It was one of the major thrills of my lifetime! Tom Nolan's paper "A Late Paleozoic Positive Area in Nevada" (1928) interpreted the rocks in the region as a single uplift. My interpretation involved the complex thrusting, folding, and uplift over a wide area of a major orogeny.

As time went on, significant details were added to the picture of the Antler Orogeny. Roberts (1964) and Roberts and Arnold (1965) described the rocks and structural features in fair detail. Later, Theodore and Blake (1975) and Theodore (2000) mapped critical areas in much greater detail, but the major features listed by Roberts and Arnold are still valid.

Today, in 2002, we know a good deal more about the orogeny than in 1949, and a redefinition follows: Antler Orogeny: a late Paleozoic orogeny named after Antler Peak, at Battle Mountain, Nevada (Roberts, 1949). Lower Paleozoic rocks deformed during this orogeny

FIG. 4.6. *Ted G. Theodore, Queen Lode barite mine, 1999.*

here are unconformably overlain by Lower Pennsylvanian strata. However, in Eureka County, Nevada, Roberts (1951) and Roberts and others (1965) recognized that the orogeny spanned from late Devonian through middle Pennsylvanian time. The orogeny consists of two major phases (Roberts and Madrid, 1990): (1) episodic early thrusting, ending in late Mississippian time, of Lower to Middle Paleozoic oceanic chert, shale, quartzite, and volcanic rocks onto coeval shallow-water carbonate rocks of the continental margin on the Roberts Mountains thrust, and (2) subsequent uplift along the Antler Orogenic Belt during latest Mississippian and Pennsylvanian time, causing clastic rocks to be shed into eastern Nevada and western Utah. The orogenic belt extends southward into southern Nevada and northward into the Mackay area, Idaho (Skipp and others, 1979), and into the Colville area, Washington (Bennett, 1937).

Ramblings in Central Nevada

On those trips into the mountains I was accompanied only by a few sparrows, meadowlarks, and shrikes, who actually flew alongside me when I was near their nests. I saw the occasional pair of red-tailed hawks or golden eagles, and mourning doves delighted in diving at me, their raids signaled only by the shrill whistle of wings. Once in a while, I would see small herds of mule deer, mostly does and fawns with a spike buck or two, and once in a while a solitary antlered buck would bound over a ridge. Marmots announced my approach with piercing whistles, and voles darted from underfoot. I seldom saw another person, except at ranches in the foothills, and that was good because I didn't really want company while learning the secrets of the rocks.

Mining During the Early 1940s

Everybody knew that the hills of Nevada had been scoured by countless prospectors, and most ore bodies exposed at the surface had been discovered. But workable ore bodies are rare and are scattered over the landscape. Aim Morhardt, a landscape painter from Bishop, California, expressed the situation well in his poem "Mining."

> The Lord was very liberal
> In spreading rocks and sand . . .
> but when He put the pay dirt in,
> He used a sparing hand!
>
> (Aim Morhardt,
> *A Thousand Acres of Nothing,* 1968, p. 12)

During the early 1940s, there was little mining in Nevada. Although gold mining had increased when President Roosevelt and Harry Hopkins in 1934 raised the price of gold from $20.67 to $35.00 an ounce, most of the moribund mines in Nevada were not affected. Nonetheless, a few placer mines were temporarily reactivated, and some kept going for a few years.

Because most lode mines, veins, and replacement deposits needed extensive rehabilitation, only a few were put into operation at this time. These few mines contained ore of fairly good grade, in the range of 0.20 to 0.50 ounce gold per ton, and they could be mined using open-pit methods. Getchell, Adelaide, Gold Acres, and Bootstrap were typical of the mines rich enough to be put into production during the 1930s.

Then, in 1942, even these mines were closed by War Production Board (WPB) Regulation L-208 for the duration of World War II. The WPB did not consider gold and silver strategic metals, and it wanted miners working to produce strategic metals and assist with other specialized tasks. Mining did continue at Copper Canyon because industry needed copper. In spite of these events, I saw a great potential for mineral deposits in the Copper Canyon and Copper Basin areas.

Fergie agreed to support me in writing my dissertation on the Antler Peak quadrangle. He told me that the ultimate decision might not be up to him but that he would do everything in his power to help me get my prize. My dream, to unravel the mysteries of the Antler Peak quadrangle, began to take shape.

But things are never simple. The Interior Committee in Congress controls the annual funding of the USGS. Therefore, all the members of the committee had to support the inclusion of money for the Metals Branch in the overall budget, and any promises made at low levels were always contingent upon Congress's providing the required funding. The money then had to be distributed by the Department of the Interior. That did not mean that Congress and the department specifically approved my dissertation area, but the copper-gold deposits at Copper Canyon and Copper Basin in Battle Mountain could be studied if the funds were appropriated for this purpose in Metals Branch funding.

In my treks around Galena in the summer of 1941, I had noted the remarkable lithology of the conglomerate later named the Battle Formation. A review of the early mining history revealed that the site of Galena had been chosen by silver miners who had discovered rich silver ore shoots in the conglomerate during the 1870s. Sadly, the ore had played out at shallow depths, but a few mines had been particularly profitable for a number of years. This metallization was very significant to me for it presaged more rich ore somewhere.

With mounting excitement, I traced this conglomerate for about three miles to the south, noting prospect pits every so often, but no extensive workings. This indicated that metallization continued, at least sporadically, to the Copper Canyon copper-gold mine being operated by the Anaconda Copper Company.

In accordance with mining protocol, I went directly to the mine office and introduced myself to the geologists there. The two men, project manager Robert Moehlman and John Collins, cordially invited me to lunch. During the meal in the mess hall, I mentioned that I was mapping the entire fifteen-minute quadrangle for the USGS and that I had studied the mines at Galena and along the ridge to the south of Galena. I also mentioned that I had a hunch where the ore bodies in Copper Canyon might be within the stratigraphic column. I asked Moehlman if I could go underground and see the ore body. He told me that he would have to forward my request to the company's Salt Lake City office. I offered Collins a quick trip to Galena to show him the silver deposits, which he accepted.

A few days later, I received my answer: a resounding NO! Sometime later when I was in Salt Lake City, I went to the Anaconda office and talked with Tom Lyon, the man in charge of the Copper Canyon project. He listened to my story but replied that the project is "high Anaconda security, and we do not want to discuss it with anyone. If two people know a secret, then it is no longer a secret! The answer is, you cannot go underground." So, that was that, but I knew that the war would end one day, and Anaconda would leave the Copper Canyon mine. I continued reconnaissance forays around Galena for a few more days then returned to Winnemucca.

Anaconda's rebuff not only failed to deter me, it fueled my interest for the job at hand, which was working out details of rock relationships, ore controls, and position of ore deposits in the regional picture. This was the story of the Antler Peak project in the fall of 1941. I felt that I had found what I was looking for—copper, gold, and silver and how they might be related to the overall regional geology of the region. I knew then what I wanted to do in geology—I wanted to look for ore deposits, such as base and precious metal deposits—in the Sonoma Range quadrangle.

First I had to go to Costa Rica, Central America, to look at manganese deposits in the spring of 1942. This project involved a good deal of travel by horseback. This was just the sort of thing that I had been trained to do in my earlier work with the USGS in Idaho, Arizona, and Washington State. I returned to the United States in May 1942 to write my report—and plan my wedding to Arleda Allen.

FIG. 4.7. *Author on horseback, Guanacaste Province, Costa Rica, 1942.*

I had been in the field in Nevada during the summer of 1941 when, much to my surprise, a letter arrived from Arleda telling me that she had returned Warren's ring and that she would be interested in seeing me again. I couldn't believe it! We began a steady correspondence, but several months went by before we got together. I was busy working in Nevada and other places. Finally we met again in the fall of 1941 in Helena, where Arleda was still teaching. On the second night of my visit, I took her to dinner, and before we had even finished dessert I proposed to her. She accepted immediately; we were engaged on our very first date.

We wanted to get married right away, but I had no idea what my next USGS assignment would be if war were declared and suggested a little delay. Arleda also felt a commitment to fulfill her year's teaching contract and so, with resignation, we set a date for the coming summer. After several blissful days together, we reluctantly went our separate ways. I knew I was hopelessly in love. Fortunately, we were able to reunite for the 1941 Christmas season. My cousin Myrtle Cook and her husband, Roy, invited us to spend the holidays with them in Newark, New Jersey. Arleda made the long cross-country train trip, and we spent our break going to plays and seeing the sights of New York.

FIG. 4.8. *Author and wife, Arleda, on honeymoon in Charlottesville, Virginia, 1942.*

After the holidays, Arleda returned to her teaching job, and I sailed on a United Fruit Company boat to Central America to begin my assignment. I had finished mapping Costa Rican manganese deposits in a couple of months, and I returned to Washington to write my report.

As summer approached, I began to make preparations for the wedding. Controls were tight, and I couldn't get permission to go west, so Arleda had to come to Washington. Mary Hewett, Foster's wife, offered to have the wedding at their home in Cleveland Park, which was quite close to the Washington Cathedral, but I had to find a minister. After some difficulty, I found Rev. Paul Yinger of the Congregational Church. I thought he had just the right kind of enthusiasm.

When the day came for Arleda to arrive, I met her plane at the airport, and she wasn't on it. I panicked. Then I learned that a general with military priority had bumped her from the airplane in Pittsburgh, and she had been forced to finish the journey by train. Hours later, at the train station, I spotted her limping down the railroad track to greet me. She had lost the heel off one of her shoes in the ordeal! I'll never forget that sight or the way we cried, laughed, and hugged. I always say, "All's well that ends well!"

Fig. 4.9. *Author and wife, Arleda, in the field, Nevada, summer 1942.*

So, on July 25, 1942, we were married at the Hewett home. Mary and Foster invited all my colleagues, including the USGS director, W. C. Mendenhall, and the chief geologist, G. F. Loughlin, and their wives. Russell Wayland, my roommate, acted as best man. It was sunny and warm, a perfect day for a wedding. After the ceremony and required cutting of the cake, some of our friends drove us to the train station. We traveled to Charlottesville, Virginia, for a brief honeymoon at the beautiful Farmington Country Club. There, for the first time in our long-distance relationship, we were alone. We savored the time, the excellent food in the dining room, the leisurely bicycle rides through the countryside, and the visit to Monticello, Thomas Jefferson's home. I knew, without any doubt, that I had found the perfect partner.

Once back in Washington, Arleda tired of sitting around an apartment while I was at work, and she set out to find a job. She quickly landed one with a firm dealing in forest products. She worked there until September, when we packed a few essentials and headed to Nevada so I could continue work on a quicksilver project in the Ivanhoe district, about fifty miles north of Battle Mountain.

This was an adventure to us. We drove for miles through the

desert, passing only sagebrush, until we reached the Butte Mine. Bob Morris, the mine manager, greeted us and showed us where we could pitch our tent. The miners greeted us warmly but without much ado. We settled into the life of a mining camp quite easily, even though, as newlyweds, we were a novelty. In fact, Arleda was one of the few women for miles around. We ate our meals with the miners in the mess tent. The food was simple but hearty. Although some may think it odd to take a new bride into such rugged terrain to live in a tent alongside a bunch of crusty miners, I will point out that Arleda was no shrinking violet. She had lived many summers in the Little Rockies, riding horses into steep mountainous terrain, and she had developed a muscular physique, so she was as at home in my world, the hills of Nevada, as she was in Washington, D.C.

At the end of the field season, Arleda and I returned to Washington, D.C., and immediately began packing for my new assignment in Central America.

Chapter 5

Mineral Deposits of Central America

All's well that ends well.

—SHAKESPEARE

Foster Hewett's strategic minerals program was extended to Latin America during the early 1940s. In late 1942 I was asked to go to Central America to work under the Board of Economic Warfare, one of the new units in the Department of State, to procure certain strategic materials for the war effort. William Schmidt, a mining engineer, Frank Simons, a junior geologist, and I were first sent to Panama to evaluate deposits of manganese, an element used in iron metallurgy. As Panama was in the war zone, Arleda arranged to go to San Jose, Costa Rica, and wait there for me.

Bill and I had read sketchy reports on manganese deposits in Mandinga Bay, San Blas Islands, and wanted to determine their potential. We flew to Panama City, where we met Frank and obtained a native guide, then we all traveled to the site. The deposits were promising enough to require more detailed study, and Frank was assigned to stay on to map them as well as look for manganese elsewhere in Panama.

Before Bill and I could leave the country, the airport was closed while part of the British fleet went through the canal on its way to join U.S. forces in the Pacific Ocean. The postponement of our flight seemed trivial. We watched with awe the parade of battleships, cruisers, and destroyers. Klaxons sounded, strings of brightly colored flags

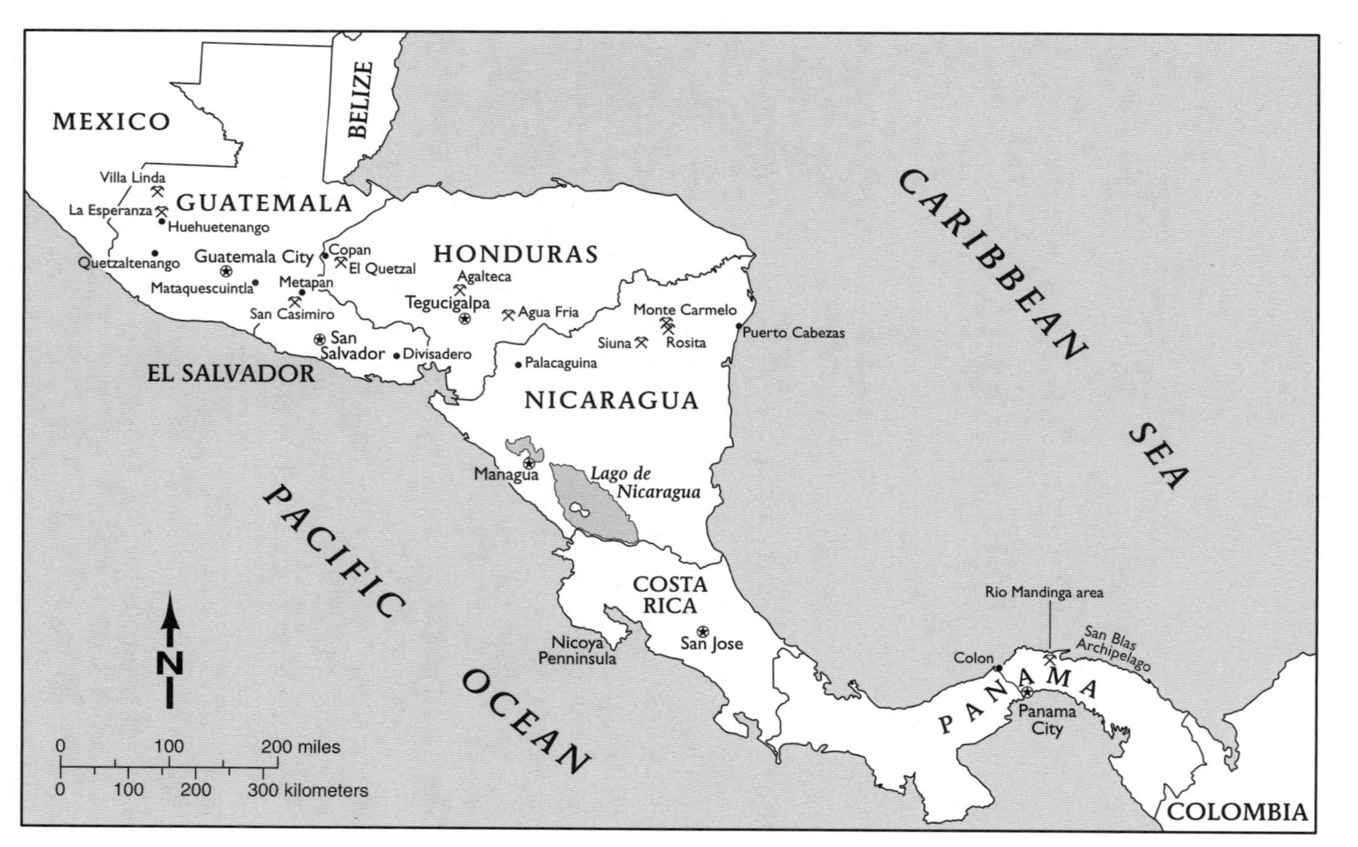

MAP 5.1. *Central America.*

fluttered, and messages flashed from semaphore mirrors as the ships passed through the canal.

We were able to travel the next day, Christmas Eve, to San Jose. I was overjoyed to see Arleda, who had taken a room for us in the Europa Hotel. She had found a tiny Christmas tree in the market and attempted to decorate it. She had tried to buy popcorn for stringing but did not know the Spanish word for "popcorn" (*palomito*). So she had gone from store to store saying "maiz poof," but either nobody understood her, or they didn't have any popcorn.

A couple of days after we returned from Panama, Bill and I traveled to the Nicoya Peninsula in western Costa Rica, along the Pacific Ocean, where I had worked the previous year. We wanted to evaluate the deposits there for future production. Getting to the deposits proved difficult, and we concluded that, given the lack of infrastructure, it would not be feasible to mine and process the ore. Later, we looked at copper-gold and antimony deposits in eastern and northern Nicaragua. On one trip we visited La Luz Mine near Siuna, about ninety-five miles west of Puerto Cabezas. We also visited Rosita, a potentially large copper-gold deposit, near the village of Monte Carmelo. Copper and gold were not strategic metals, but copper is an important industrial metal. I would have loved to spend a month or two at each place, for the copper-gold deposits were very similar to those at Copper Canyon, Nevada, but we moved on to search for strategic minerals.

We visited antimony deposits near Palacaguina, but the deposits proved too small to put into production. Then we traveled to El Salvador, where we made our temporary headquarters in San Salvador. We traveled south to the gold-silver mining district near San Sebastián, Divisadero, and then northeast to a lead-zinc district near Metapan. Neither district contained strategic mineral deposits, so we left San Salvador for Tegucigalpa, Honduras.

There we settled in a pensione for travelers run by TACA (Transportes Aereas Centro-Americanos) airlines. We started with the Rosario Mine, where Bill Schmidt had once worked. The mine, a major silver producer from the time of the conquistadores, employed geologists who were familiar with the geology and mineral deposits of Honduras. They told us about a number of antimony and mercury deposits, and we decided that I should go later to western and central Honduras to examine them.

We then went to Guatemala City, where we joined the Board of Economic Warfare Mission (later Foreign Economic Administration) under Mission Chief Ralph Dewey. The purpose of this mission was to seek strategic materials, including chromite, antimony, tungsten, nickel, mercury, mica, and quartz, and commodities such as quinine, sisal, lemon grass, and special kinds of lumber, a rather unlikely mix, but everything played a part in the war effort. The minerals program was carried on by Bill Schmidt and me, and Earl Irving, from the Agua Fria Gold Mine, soon arrived to work with us. Earl's experience in the region made him a valuable addition to our party; moreover, Earl's wife, Julietta, was Honduran. She helped Arleda and me get around in Guatemala and often served as translator for us. Next to join our team was Lawrence Houk, a specialist in quartz and mica deposits.

Guatemala City

Arleda and I settled in Guatemala City. At first we stayed in the San Carlos Hotel with the rest of the mission personnel. But Arleda was pregnant, and we knew we would have to look for an apartment. The task of looking fell to Arleda. One day she was going through an apartment building in the downtown area with the building's supervisor when a woman came out of her apartment and, overhearing the rental figure quoted to Arleda, berated her for accepting such a high figure. She suggested that Americans didn't care what they paid, which made things difficult for everyone else. When Arleda burst into tears, saying that all she wanted was a nice home for the forthcoming baby, the woman took Arleda into her apartment and made amends by serving tea and cookies. Arleda learned that the woman was an Australian, and her daughter, Jean Chisholm, worked for the American embassy. We got to know both women, and Jean became a lifelong friend.

Michael Foster Roberts

While Arleda was apartment hunting, I, thinking that the birth of our child was a couple of weeks away, went to northwestern Guatemala, near Huehuetenango, to study ore deposits. I was staying there when our son, Michael (named after Foster Hewett), arrived. Arleda had been dancing with Earl at a party when her water broke. She was hustled off to the Presbyterian Hospital, and Michael was born on August 8, 1943. I was informed of the event by telegram and left immediately for Guatemala City, but bus travel was not quick, and it was two days before I caught my first glimpse of our son. (He was about twenty-three inches long and weighed six pounds, fifteen ounces.)

We finally found an apartment in the Mansion San Francisco, but we stayed for only a couple of months. Later we moved to Fourth Avenida Sur near the English-American School where Arleda had taken a job teaching music and English.

To keep the household running smoothly while we were both working, we hired a nursemaid and cook. Rosa, the nursemaid, was a Maya from Coban, and Luz, the cook, was from Guatemala City. They ran our house competently and did an excellent job of rearing Michael for nearly two years. They taught him Spanish as his first language, and he absorbed English by listening to Arleda and me.

Arleda's job at the English-American School gave her ample outlet for her talents. Because teachers there came from a variety of backgrounds, students could choose the accent they preferred to imitate. Arleda spoke with a northern American accent, another teacher had a broad southern drawl, and still another had a clipped British accent.

Arleda was especially successful in her role as music teacher, and she capped her career at the school by taking the girls' chorus to give a concert at the cathedral. The priests would not permit some songs in the school repertoire to be sung inside the church, so Arleda had to choose other acceptable songs, but the concert was a huge success.

Bill Schmidt, Earl Irving, and I studied the mineral resources under the overall supervision of Alan Bateman, who had been one of my professors at Yale. Earl evaluated the chromite resources in Guatemala

FIG. 5.1. *Michael, at two months, and nursemaid Rosa in Guatemala City.*

FIG. 5.2. *Luz holding Michael and flanked by Arleda, Billy Welborn (a friend), and Eva Roberts, Guatemala City, 1944.*

Fig. 5.3. *Arleda, wearing a crimson blouse (*huipil*) from Tactic, northern Guatemala, and heavy silver marriage chains around her neck.*

Fig. 5.4. *Arleda and her choir from the English-American School after a recital at Guatemala City Cathedral, 1943.*

and helped one of the producers set up a company to ship chromite to the United States. He also, along with Lawrence Houk, organized a quartz and mica project. I took on the job of looking at mercury, antimony, and manganese deposits.

Northwest Guatemala

Northwest Guatemala contained only small deposits of mercury and antimony but had a number of base metal deposits (copper, lead, and zinc) that warranted evaluation. I traveled from Guatemala City to Quetzaltenango, then north to Huehuetenango.

The metal mines in that area were mostly in the Cuchumatanes Mountains, a massif of limestone, serpentine, and conglomerate that attained an altitude of more than 14,000 feet. I had gone from city to city by bus, but I made trips into the Cuchumatanes on horseback, accompanied at times by Arleda and friends. On such trips Michael and Rosa or Luz might accompany us and stay at a nearby hotel while we rode off into the field.

Earl also accompanied me on a couple of trips to lead-zinc mines. Once we visited Villa Linda on the west slope of the Cuchumatanes. To reach the mine we had to climb to the crest of the range and then follow a canyon down. Villa Linda's miners produced small quantities of lead. They mined oxidized ore and then washed most of the iron oxides out, using a simple trough launder, which left a concentrate of mostly lead carbonate and lead sulfate. This material was mixed with charcoal and then reduced to lead in a simple furnace. The molten lead was cast in bars for sale throughout Guatemala. The operators eked out a meager living at this work but were proud of their independence. When we went into Villa Linda Mine, our carbide lights extinguished suddenly when we were about seventy-five feet inside. The oxidizing process (the change from sulfides to lead carbonates and sulfates) depleted the oxygen in the air. But the men and boys who worked the mine by flashlight demonstrated no apparent lung damage, so Earl and I continued on to map the twists and turns of the shafts as they followed the ore downward.

We had another unnerving experience in the La Esperanza Mine, near Finca Chochal, at an altitude of 9,700 feet. When Earl and I visited the mine, we found small bodies of lead carbonate ore along a limestone bed; small-scale stoping operations had been carried on in recent years, but the workings were extremely dangerous. The access drift began in a jumble of huge limestone blocks averaging ten feet across. We were able to work our way down, but the drift was so narrow that we could not turn around, and as we passed a partly caved side drift, our guide said, "Alla yace pobre Juanito" (There lies poor Juanito). We were able to turn around at the end of the drift, rather than having to back out, and once outside again we finished sketching a map of the workings. The experience left us chastened and full of empathy for men who had to work in such dangerous mines.

We also visited lead mines in other parts of Guatemala, some of which appeared to have potential (and are rumored to have since begun small-scale production), and examined copper deposits in several mining areas, including San Mateo, Mataquescuintla, Las Sandillales, Zuhoj, Cerro Vivo, and Rio Oxej.

Rio Oxej is in the chicle-producing area of the rain forest. (Chicle, a gum made from the latex of a tree, was then the main ingredient in chewing gum.) The air was humid and the heat oppressive, and the native workers wore only minimal clothing. I opted to retain my clothes, even though they were soaked with perspiration. The copper deposit had potential; the vein was a couple of feet wide and contained pyrite and chalcopyrite, copper-iron sulfide. The sulfides were locally oxidized.

Honduras

Honduras was an exciting place for us to work, as we were the third group of people to exploit its minerals. The Mayas had begun mining and smelting processes ages ago, and the conquistadores had continued after subduing the natives. In western Honduras some gold mines were associated with iron and antimony deposits. Earl and I also mapped iron deposits in central Honduras in 1944. Philip Chase, of the Oliver Iron Company, evaluated the deposits but found them too small to warrant building a railroad to the mine site.

FIG. 5.5. *Earl M. Irving and the pilot of a chartered aircraft at airport, Agalteca Iron deposit.*

One antimony mine was commonly reached via Copan, one of the famous Mayan ruins. On one trip I visited Copan along with Don Manuel Bueso (owner of a large *finca,* or ranch) and Rudolf Nater (owner of El Quetzal antimony mine). I had arranged to meet Nater at the mine on a certain day. Arleda and I traveled from the railroad in Guatemala to the Bueso Finca via Copan. We arrived at the finca on a Saturday evening, had a pleasant visit with the Bueso family, and rose early the next morning ready to go to the mine. Two guides from the finca were reluctant to start out Sunday morning, saying that Arleda's mule had to be shod. They shod the mule, and we started out in the early afternoon. It had rained heavily before we arrived, and the river (a tributary of Rio Chamelecon) was high, but the water did not reach the bellies of the mules, and we crossed the river easily, calling out "yippee-kay-yay." The path to the mine became a steep, narrow trail. As we continued climbing, the sky darkened and it began to rain softly. When we finally reached the mine, night had fallen, and to my dismay, no one was there. I felt like a fool for insisting on traveling that day.

We located the mine headquarters and found a note on the door from Nater, informing us that a shooting had taken place at the vil-

FIG. 5.6. *Before the storm, en route to the El Quetzal antimony mine. Arleda had just forded the rain-swollen river, 1944.*

lage below and that he had to officiate at the inquest. He asked us to come down to the village. This posed a real problem. By now it was raining hard, but we had no other choice, so we headed down. One guide went first on the narrow trail, then Arleda; I followed, and the second guide brought up the rear. In some places the trail had been cut through solid rock, which was not too bad, but in spots the dirt had turned to dangerously slippery mud. On one such treacherous place, Arleda's mule lost its footing and slipped off the trail. The two of them landed about twenty-five feet below on a soft debris slope. The mule rolled over Arleda and then tumbled down the hill alone. In those brief seconds I feared the worst, but incredibly, Arleda had landed on a soft soil bank, and instead of being crushed when the mule landed on her, she had only been pressed into the rain-soaked earth. Even more miraculously, her feet had come free of the stirrups when the mule rolled over her before careening down the hill. I called to her anxiously, "Arleda? Are you hurt?" Of course, that was a stupid question—how could she avoid being hurt with the mule falling on her? I leaped over the edge and slid down to where she was lying. Much to our relief, she suffered no broken bones. She was badly bruised but otherwise uninjured. She later said that the only thing in her mind was "Will I ever see my baby Michael again?" The guides helped me get her back on the trail; we conferred briefly and decided to walk down the hill and lead my mule. We found her mule, which had landed upside down between two trees, and had only gotten a little scratched. Our guides righted the mule but left her tethered to a tree as the slope was too steep to climb back up. They said that they would return for her in the morning. Arleda and I continued down the trail, slipping and sliding on the steeper portions, until we finally reached the village, nearly two miles below the site of the accident.

When Mrs. Nater heard about the accident, she immediately awakened some servants and put them to squeezing lemons and collecting the juice in a washtub. She told Arleda to sit in it, saying that a lemon juice bath was the best treatment for bruises. Surprisingly, the treatment worked. We rested four days at the Naters'. Arleda recuperated, and the mule was recovered and brought to the village; the fall had not harmed her in the least. When the rain stopped, I went up to the mine, conducted my examination, and returned to the village. We learned later that there had been a major hurricane in the Caribbean, and our storm had been part of the aftermath. On the fifth day, Arleda felt well enough to ride again. At Esquipulas, Guatemala, we exchanged our mules for a bus to the train station, where we caught a train back to Guatemala City. The guides returned the mules to the Finca Bueso. It was another case of "all's well that ends well!"

It gradually became obvious that Central America would not yield major amounts of either strategic minerals or base metals, and our project was expanded to include compiling a new geologic map of Central America. I had also planned to join with Bill, Earl, and Frank in writing a report on the mineral deposits of Central America. But in 1943 Bill decided to return to his retirement home on Long Island and left the remaining work on mineral resources to the three of us. Earl left the project in 1945 to join a USGS project in the Philippines, where he worked for several years before going on to a project in Colombia. I sandwiched the completion of the report in between other work, and it finally came out in 1957. While not as polished and complete as I would have liked, it did contain much information on the mineral deposits of Central America, which had not been well documented since gold and silver deposits had been discovered and mined by the early Spanish conquistadores. Operations since then have been intermittent and on a small scale. (Since the report appeared, some oxidized nickel deposits containing the mineral garnierite, nickel silicate, have been found in ultramafic rocks in Guatemala. These deposits were mined in the 1960s.)

In May 1945, Arleda, Michael, and I returned to the United States. We found an America much changed but jumped in and reorganized our lives. I went into the Military Geology unit of the USGS, while Arleda obtained a position with the Department of Commerce.

Chapter 6

Military Geology

> War . . . is all Hell.
>
> —WILLIAM TECUMSEH SHERMAN

The USGS was deeply involved in World War II. Survey geologists served as technical advisors, provided special reports and maps, and worked as consultants to field operations.

Early on, the question was raised: Should USGS geologists be commissioned as officers in the various service branches that needed geologic help or advice, or should they be retained as liaisons? Sidney Paige, U.S. Army Corps of Engineers, a former commissioned officer in the U.S. Army, offered to look into this problem. The armed forces eventually decided that it would be simpler for the USGS to furnish specialists who would act as liaisons when and where they were needed. Some men were sent overseas as scientific consultants, but most were used in the United States.

USGS geologists carried on a variety of investigations. For example, Kenneth Lohman, a USGS micropaleontologist, was asked to identify the source of ballast sands that had been carried by balloons released from Japanese submarines and retrieved near our Pacific Coast. Some of these balloons carried incendiary devices, designed to set forest fires, and others carried explosives intended to endanger or frighten residents who found them. The coastal forests were generally too wet to ignite easily, and the sparse coastal population did not scare easily. Although the balloons were never a serious threat, our military wanted to know where they had been assembled. It turned out that the ballast sand could be identified by microscopic Foraminifera (single-celled marine organisms) as coming from a certain beach in Japan. Had military leaders so chosen, they could have bombed the beach to destroy the factory involved. John McPhee, a writer who has produced many books and articles on the geology and environment of the American West, wrote a *New Yorker* article (January 29, 1996) about other unsung USGS heroes who worked quietly and anonymously behind the scenes to help the war effort.

It was inconvenient to be assigned to Military Geology immediately on arriving in the United States, because I was attempting to write the report on my two years in Central America. However, this made no difference to the branch chief. It turned out that several geologists who had been working in Military Geology had been assigned as scientific consultants to General MacArthur in the Philippines. So help was needed on several ongoing projects.

Military Geology headquarters were in the basement of the Old Interior Building (nicknamed "the dungeon"), where security and privacy could be easily maintained. I joined a group working under E. S. Larsen III. This group was small, averaging no more than eight scientists. Mark Pangborn of the USGS library staff was assigned to us as a librarian. When a new project came our way, he would quickly assemble all the pertinent reference maps and reports.

Our group tackled two principal projects. One was an urgent problem in testing the SCR-625 mine detector, which had not performed well in southern Italy and Sicily or in the Pacific Islands. We were charged with finding the reason. The other chore was the preparation of folios and reports for field operations of the armed forces (strategic engineering studies) as well as intelligence reports (JANIS, or Joint Army-Navy Intelligence Service).

Mine Detector

My first assignment was helping evaluate the SCR-625. Mine detectors (metal detectors) of that time consisted of two coils of wire that were aligned at right angles to each other. When the coils were electrified, if a metallic conductor or magnetic body was brought into close proximity to the coils, a current was set up that could activate either an ammeter (a device to measure electrical current) or a beeper. For some reason, the SCR-625 detector was not doing what it had been designed to do. We set to work preparing a facility for testing metallic mine detectors. The project had been started by Konrad Krauskopf (Stanford University), Edward Sampson (Princeton University), and M. M. ("Red") Striker (U.S. Department of Agriculture). Ed Sampson was preparing to accompany U.S. troops who were returning to the Philippines with General MacArthur. He and Konrad gave us a quick rundown on the project status, having determined that the mineral likely to be the culprit was a magnetic iron oxide (such as maghemite or gamma-hematite). Maghemite was capable of affecting a mine detector like magnetite or metallic iron. So, Red and I set out to learn all we could about where and how maghemite formed. Konrad headed back to Stanford, and Ed went to the Philippines.

Red had access to soils from all over the world, and he selected about twenty-five thousand that came from climatic zones that had proved troublesome for the SCR-625 mine detector. Before we could begin tests, we had to have a device built that could rapidly measure the magnetic susceptibility (response) of soils. We posed this problem to R. J. Duffin who studied terrestrial magnetism at the Carnegie Institution of Washington. He quickly built a model of a mine detector into which a test tube of soil could be inserted for direct measurement. We calibrated it with manganese sulfate, which has a low, but precise, magnetic susceptibility, and were in business. After running a few hundred samples of soils from critical areas, we had an answer. Soils with a little iron in the ferrous state could be oxidized to maghemite and, thus, cause a signal or current in the detector.

We found that soils with the highest magnetic susceptibility were formed in tropical to semitropical regions derived from iron-bearing rocks, such as volcanic rocks with a high iron content. The U.S. Army Corps of Engineers' research laboratories at Fort Belvoir, Virginia, assigned R. E. (Bob) Stewart and a small group of men to work with us. They quickly realized that a more sensitive mine detector was needed and so began to design one. They asked us to supervise preparation of a number of lanes about 150 feet long, 2.5 feet wide, and about 8 inches deep, each filled with soils of different textures (clay, loam, sand) and with different magnetic susceptibilities (mixed with varying amounts of titaniferous magnetite). These lanes proved useful in development of new detectors. Ed, Red, Bob, and I published a joint report on the project in 1949.

Military Folios and JANIS *Reports*

Our work in Military Geology included the preparation of folios designed to assist our forces in an invasion of Japan's home islands. The folio on the island of Kyushu had already been prepared, and so I worked with a group assigned to the main island, Honshu, which included Kobe, Tokyo, and Hiroshima. The folios consisted of maps prepared for special purposes and containing such details as the locations of water supplies, trafficability for tanks and trucks, possible airfield sites, and other information that might be helpful to advancing armed forces. The water supply map, for example, identified the location of wells of potable water because the surface water supply in Japan is polluted by human wastes used in fertilization of crops. A source of pure water was essential. We were given 1:50,000-scale maps (about 4,000 feet to the inch) of the region for this purpose. A translator from the U.S. Army showed us the Japanese character for *well.* We perused the maps of the area, happily circling all of the deep well symbols. When the translator came back a few days later to check our work, he burst into laughter; we had found most of the breweries on Honshu.

A trafficability map was necessary because invading forces would need to know along which routes tanks and armored vehicles could advance. We commonly used the topographic and geologic maps to identify suitable and unsuitable areas.

Sites for airfields could also be plotted on a modified geologic and topographic map. The work on the preparation of folios was intense and required long hours; we often worked late into the night in order to meet deadlines. In between my work on the folio, I was assigned to write JANIS reports on areas for which the armed services needed summary reports on the geology and mineral resources. The military seemed to have an insatiable appetite for such material.

I worked on two JANIS reports, one on Argentina, because I spoke and wrote fairly good Spanish after having been immersed in it in Central America, and another on Szechwan, China. We were all very glad to be advised to suspend our work on the invasion folios after the atomic bomb was dropped on Hiroshima, August 6, 1945. The second bomb was dropped on Nagasaki on August 9, and thereafter Emperor Hirohito surrendered. Although we were not military strategists, our knowledge of the terrain of the Tokyo area told us that taking the home island would have meant enormous losses among our troops. We didn't feel any guilt about the dropping of the bomb, knowing of Pearl Harbor, the Death March at Bataan in the Philippines, and Japanese atrocities in Burma and Singapore. By dropping the bomb, we surely saved many American and Japanese lives.

When the war in the Pacific was over, our unit was disbanded. Some units were sent to Japan to help revive that country's economy; others carried on geologic studies of the Pacific Islands. The rest of us went back to our normal projects.

Chapter 7

Postwar Washington, D.C., and the Defense Minerals Exploration Program

SALT LAKE CITY USGS, 1950–1954

This is the place!

—BRIGHAM YOUNG, 1847

After World War II ended in the fall of 1945, I was transferred back to the Metals Branch, which was then being run by Olaf Rove. Although I was itching to get back into the field, Olaf asked me to help reorganize the branch's manuscript processing unit. I would be allowed a few months away each summer so that I could finally complete fieldwork and write a report on my Antler Peak project. Also, I would be able to work on my part of the Sonoma Range reports, which had been shelved during World War II. And, in addition, I would have to find time to finish my dissertation for Yale if I were to make the ten-year deadline.

As head of the Manuscript Processing Unit (MPU) I was responsible for the flow of reports through the branch. Each manuscript went through four stages: a technical critic reviewed it; specialists checked the geologic and geographic names used; editors reviewed it for style; and the chief geologist and director's office gave it final approval. I liked the job because my being assigned it meant that we were returning to prewar USGS priorities. I received an in-grade promotion for my effort. It wasn't much extra in my paycheck, but it was a plus on my record for a job competently done.

While I was running the MPU I also worked with Julian (J. D.) Sears on several new series of maps. It seemed that the prewar emphasis on USGS bulletins and professional papers had resulted in slow publication of our data, so J. D. proposed that we set up map series for specific purposes. He and I devised a program for the development and publication of several types of maps: a mineral resource series, oil and gas maps, and other specialized maps. The existence of these series permitted rapid release of important new data through updated editions of maps. (The series are still used today.)

I worked at MPU for two years. When that work and work on the new map series were finished, I would finally be free to head west. All I needed was an assignment to get me back there. Anticipating the time when I would be done with the Antler Peak reports, I approached Olaf and asked him if I could take on the Bingham Copper-Gold project. He said, "I'll make you a deal, Ralph. You go out to Salt Lake City and head up the new office for the Defense Minerals Exploration program for a few years and then you can tackle the Bingham project." I agreed, knowing that this would be the best way to get back to the West and the field.

Michael, Steven, and Kim

When we had arrived in Washington, D.C., in mid-1945, Michael had spoken mainly Spanish. Luz Bautista had come to the United States to work for us shortly after we arrived, thus freeing Arleda to work for the Department of Commerce. Michael continued to speak Spanish with us and Luz, then one fateful day, some playmates accused him of being Chinese. He did not know exactly what that meant, but he decided to speak English from then on. I discovered this when he came up to me and asked, "Whatcha doing Daddy?" I was assembling a bookcase, and I told him so, in English. From that point on Spanish was his *second* language.

While I was finishing my work with MPU, my family was growing. Our second son, Steven Arland (named after Arleda's brother), was born on March 2, 1947, and our last son, Kim Halcot (named after my father), on July 27, 1949. Michael was overjoyed by the arrival of his

FIG. 7.1. *Steven at one year old and author, Copper Canyon Mine, March 1948.*

brothers, and Arleda and I felt that three boys suited us just fine. Steven was named for Stephen R. Capps, the supervisor of my second USGS 1936 project in the Dixie district of Idaho. Michael gave Steven the nickname "Chico," "porque era más pequeno" (because he was smaller), which stuck for many years.

In late 1947 we took a break from Washington, D.C., and returned to Antler Peak so I could finish the underground mapping at Copper Canyon. We spent the winter of 1947–1948 there, and Steven celebrated his first birthday at Copper Canyon.

Michael and Steven loved life in the field. They played in the mines and shacks and looked for minerals on their walks over the rocky hillsides. I showed them how to make rockets using the fuel for miners' lamps, and we used to blast tin cans high into the air. During that winter, things got serious when they both contracted scarlet fever. Fortunately, there were sulfa drugs to treat the illness and bring down the fever, and neither boy suffered brain damage nor any of the other long-term effects that had occasionally accompanied the disease in the days before sulfa was available. During the quarantine period, Arleda and the boys took over the staff house at the mine while I carried on alone at our house.

FIG. 7.2. *The Roberts boys during a light moment, 1949.*

Kim had health problems as an infant. We were not sure of the reasons but suspected that he might be allergic to environmental contaminants in Washington, D.C. We decided that a change in surroundings might help his condition, and Fergie suggested we move out to his farm during the summer of 1949. Kim's condition improved immediately. The next year, we were able to leave Washington entirely.

Moving to Salt Lake City, Utah

In November 1950, Arleda and I packed up our household effects, arranged to sell our condominium, and took off for Salt Lake City with Michael, Steven, and Kim in tow. Jean Chisholm Lingham, our old friend from Guatemala (who had recently married John Lingham, a British army brigadier), came along to help Arleda with the driving. They, along with Steven and Kim, traveled in our 1949 Mercury station wagon. Mike and I followed in a tired but dependable 1938 Dodge.

In mid-November bad weather was a possibility no matter how we went west, but we chose a southern route, hoping to at least avoid any major snowstorms. We planned to go by way of the Cumberland Plateau, over to Memphis, through Oklahoma City, down to Amarillo and Albuquerque, and then up to Salt Lake City. Our objective was to get as far south as we could as quickly as possible. There were signs of trouble from the start. As much as we had tried to be organized, we were still late getting on the road. The first day we made it through the Appalachian Mountains and into the middle of the Cumberland Plateau. Finally, in a town somewhere between Knoxville and Nashville, we found a small motel where we could spend the night. (In those days there were no interstate highways or national hotel chains.) We turned in, and when we woke the next morning, we were in a snowy winter wonderland. Much as we'd tried to avoid the bad weather, we'd landed right in the middle of it.

I decided we couldn't delay; we'd have to push on and hope for better weather. When I drove the Mercury to a service station at the bottom of a hill, the car spun out of control and "swapped ends." Fortunately no other cars were near. Later, after I'd gotten both cars refueled, we inched our way carefully across the plateau. We descended into the Mississippi Valley, and we were clear of the storm. We proceeded west through Arkansas, Oklahoma, and Texas. We had reached Amarillo and were thinking the rest of the trip would be smooth when the water pump gave out in the Dodge. I spent all night replacing it, caught a couple hours of sleep, and we started out again, only to find that the Dodge's fuel pump had broken. Gas was not leaking—it was *pouring* out of the fuel pump. I knew my limits, so I found a garage with a mechanic and took the family and Jean to a café for breakfast while we waited for repairs to be completed. Then, fingers crossed, we headed for Albuquerque.

A slight detour through Holbrook, Arizona, allowed us to drop off Michael and Steven with my parents, who planned to take them back to Van Nuys, California, for a visit. Arleda and I knew getting settled in Utah would be enough of a challenge without two very energetic young boys underfoot. We continued on to Salt Lake City.

When we arrived in Utah, we found a house in Murray, some miles outside Salt Lake City. It was a large four-bedroom house that had been built for the superintendent of the American Smelting and Refining Company plant. We were just south of the "bag" house of the old smelter where volatile constituents, such as arsenic, antimony, and bismuth oxides, had been collected. The real estate agent who sold us the property recommended that we not plant a garden containing carrots, radishes, and the like, because the soil in our backyard was very toxic. In retrospect, it seems to have been a dangerous place to live, but we didn't eat any homegrown vegetables, and we seem to have suffered no ill effects from our three years in the house.

Kim's allergies, so bad in Washington, D.C., disappeared in the dry air of Utah. We were glad we had made the move and found much to like about the house and its surroundings. The boys played in the backyard and on the grounds of the abandoned smelter, exploring caves and storage sheds. They also created "forts" in some of the unused rooms of the upper stories of the house. They could play there any time of year because the house was built with double-brick-insulated walls. Michael and Steven were devoted to Kim and protective of him, although they all enjoyed good-natured roughhousing. Early on Kim showed an interest in electricity. When he was only three we began to notice that he took particular enjoyment from lugging a space heater around, plugging it into one outlet, leaving it for a while, then unplugging it and moving to another outlet. He did this every day for several weeks and all of a sudden stopped. I asked him if he had received a shock, and he solemnly nodded. I explained to him that if he touched the terminals of the plug when he put them in the outlet he would get a shock. This early interest in electricity presaged a career in electronics.

Kim also relished outdoor play with his brothers and me. I had been teaching Mike and Steve to play baseball, and Kim was just as eager to learn. When he was learning to play ball, I would pitch a ball for him to bat. I will never forget his frustration as he would try, and fail, to bat the ball back to me, and I remember the radiant smile that transformed his face when he at last connected with the ball. Later Kim was a powerhouse in Little League baseball.

Defense Minerals

By March 1951, I had the defense minerals exploration office running smoothly, in no small part due to the excellent staff I was able to put together. Roscoe Smith and William Hasler, both experienced mining men, ran the program along with Robert Osterstock and several bright young geologists who had just finished their university training. In a short time we had a steady stream of customers pouring into the office, mostly prospectors who needed help testing their properties for base metals and strategic minerals. We were quite busy over the next three years, but, whenever possible, I went out to look at the geology of areas new to me in western Utah and especially eastern Nevada, and I took the boys with me as often as I could.

Pointing for Eureka County, Nevada

In the early 1950s two big changes in the USGS led us in a new direction. The defense minerals workload was lessening, and Charles Anderson, new chief of the Metals Branch, decided to merge the functions of the Salt Lake City offices with those of the California offices in Menlo Park, about thirty-five miles south of San Francisco. At about the same time, Arthur E. Granger, chief of the Eureka County project in north-central Nevada, caught the uranium fever and decided to resign and join the uranium seekers. Thus, in early 1954, the Eureka County project was left leaderless. I was offered the job, and I grabbed it without hesitation. Finally, I could get back to mapping the geology of Nevada in Eureka County. Even though home and office were now in Menlo Park, we went out to the field in Nevada each summer.

Chapter 8

Eureka County

1954–1965

Aurum irrepertum et sic melius situm
(Gold undiscovered and all the better for it being so).

—HORACE

Any geologist or prospector could have discovered a major gold deposit in the Carlin Belt prior to the 1960s, but I am glad that the discovery waited until I and Newmont Mining Company geologists came along!

Thanks to the uranium boom, the geological mapping project in Eureka County was mine. I was once again in the right place at the right time. I had a once-in-a-lifetime opportunity, and the Bingham copper-gold project could wait on the back burner for a while! I was now responsible for the cooperative project between the USGS and the Nevada Bureau of Mines and Geology. I could map the critical area, tracing the Roberts Mountains thrust throughout the county and showing the gold deposits in a simple model. The undertaking promised a scientific and economic bonanza!

Eureka County was a critical place for me during the 1950s, as it was to lead me to a better understanding of the geologic framework of northern Nevada. I had a good knowledge of the geology of the Sonoma Range quadrangle and especially of the Antler Peak area, but there were still great gaps remaining. The least understood zone extended from Carlin to Eureka via the Roberts Mountains. In my earlier forays into central Nevada I had been aware that oceanic chert, shale, and quartzite existed there. The rocks were similar in lithology to the Paleozoic rocks we had mapped in the Sonoma Range quadrangle. What did this mean to me? Read on!

Let us look at the history of geologic studies in Nevada. The first geologists to study the rocks in north-central Nevada were associated with regional surveys in the late 1800s, such as the Fortieth Parallel Survey under King. These geologists mapped the major elements of the geology from the Rocky Mountains to California, but the project was so vast it did not permit attention to detail; moreover, many critical aspects of modern geology were unknown at that time. Today, we would call their work a broadbrush, or reconnaissance, survey. These early geologists were quite at home in the Rocky Mountains and followed geologic units with which they were familiar into western Utah and eastern Nevada. When they reached a point about fifteen miles west of Elko, they encountered a great deal of chert and shale, which had never before been mapped in that area. They failed to distinguish between interlayered chert and shale and younger conglomerate, sandstone and shale composed of chert and shale fragments. The chert and shale later proved to be of Ordovician through Devonian age, and the conglomerate, of Mississippian age, resting unconformably upon the chert and shale. The early geologists correctly inferred that a shoreline existed to the west, but they could not take the time to map the rock units in a consistent manner. In the Antler Peak area they correctly identified fossils in the Antler Peak limestone as Upper Coal Measures (now Pennsylvanian) but grouped the underlying conglomerate and related rocks with older Paleozoic rocks of the Roberts Mountains allochthon. Consequently, they failed to identify the key to the Antler Orogeny.

Another group of geologists, under Arnold Hague, made detailed studies of the stratigraphic sections in the Eureka, Nevada, mining district, about ninety miles south of the area covered by the King survey at Carlin. In 1892, Hague and his associates mapped rocks similar to those encountered by the King survey but were able to date the conglomerate as Lower Coal Measures (Mississippian)—late Paleozoic in age. Hague also correctly interpreted the conglomerate as indicating orogeny with the main zone of uplift to the west.

During the 1920s, a new breed of geologists came along to study the rocks and ore deposits of Nevada. These men mapped smaller areas than the regional surveys, and consequently, they could pay more attention to details of the rock units and of structures that deformed the rocks. Henry Ferguson, my mentor, for example, mapped the Manhattan district in central Nevada (1924), where he recognized Paleozoic deformation, including folding and thrust faulting of chert and shale over limestone. He also recognized that these rocks and structures controlled the ore deposits.

What Is It, a "Positive Area" or an "Orogeny"?

In 1928 Tom Nolan, at Gold Hill in western Utah, compiled a record of the regional geology of western Utah, Nevada, and northeastern California and concluded that Ferguson's late Paleozoic deformation was merely part of a broad "late Paleozoic positive area," correlating rocks across this region. This work assembled useful data, but it did not build on the work of King, Hague, and Ferguson. Nolan apparently assumed that the rock units of western Utah continued across Nevada into California with only minor changes. This assumption was proved erroneous by the work of Edwin Kirk, Charles Merriam, and Charles Anderson.

In 1938 Nolan moved to Eureka, Nevada, to study the geology and ore deposits there. The Eureka area contained a magnificent section of carbonate rocks that ranged from Cambrian through Devonian age as well as an excellent section of late Paleozoic rocks, including thick units of Mississippian clastic rocks made up of conglomerate, sandstone, and shale. The conglomerate consisted largely of chert, quartzite, and shale boulders, cobbles, and pebbles. These rocks had been studied in fair detail by Hague, Emmons, and others in 1892; however, Tom Nolan wanted to remap this area.

Early on Nolan invited Charles Merriam, a paleontologist from the University of California at Berkeley, who specialized in Devonian faunas, to work with him on regional geologic problems. In company with Charles Anderson, also from Berkeley, Merriam worked on a problem that Edwin Kirk (1933) had unearthed in the Roberts Mountains about forty-five miles northwest of Eureka. Here two distinct facies of Paleozoic rocks, carbonate rocks of shallow-water facies (Cambrian through Devonian age) and partly age-equivalent siliceous rocks, mostly chert and shale, of deep-water facies were juxtaposed. Kirk did not determine the relationship of the two facies. The solution of this problem would prove to be critical to understanding the geology of north-central Nevada. Merriam and Anderson (1942) found that these two facies were separated by a major thrust fault that they named the Roberts Mountains thrust and dated it tentatively as late Cretaceous or early Tertiary in age (about eighty million years ago).

Sonoma Range

Meanwhile, in 1939, Fergie, Si, and I had begun work in the Sonoma Range area, and we found Nolan's simple "late Paleozoic positive area" in reality to be a fantastic, full-blown major orogeny, comparable to the Appalachian Orogeny of late Paleozoic age in eastern North America. As I had the best exposure of the orogenic features in the Antler Peak area, I named it the Antler Orogeny in my dissertation at Yale in June 1949, and in November of that year, I announced it at the annual meeting of the Geological Society of America (GSA) in El Paso, Texas.

Roberts Mountains

Being a part of the Roberts family, I was interested in the origin of the name of Roberts Mountains. I found that they were named for Bolivar Roberts, superintendent of the Pony Express mail route from Salt Lake City to Sacramento in 1860–1861. The Roberts family has long been involved in mining, first in Wales, then in the United States in Colorado, Nevada, and California. Bolivar Roberts first gave his name to the Pony Express station on Roberts Creek, then to Roberts Creek Mountain, the highest peak in the range, and finally to the Roberts Mountains. (Wondering if I might be related to Bolivar Roberts, I asked a nephew, Fred Norton, a member of the Mormon Church and a son of my sister Louise, to look up our genealogical connection, going back as far as necessary. Lo and behold, he found a connection, back in the eighteenth century, not on my father's side, but on my mother's side through the Hurlburt family! This was not what I had expected, but it satisfied my curiosity. So now, back to geology.)

Robert E. Lehner and I began mapping Eureka County in 1954, first by tracing the Roberts Mountains thrust northward from the type locality through the Simpson Park, Cortez, Shoshone, and Tuscarora Ranges, proving that the thrust had regional extent. We were able to prove continuity by mapping identical relationships of the thrust plate to lower carbonate rocks from range to range. Moreover, the thrust plate was overlain by mid-Mississippian rocks near Elko, Nevada; Eureka; and other localities, which proved that the thrust predated mid-Mississippian time (about 330 million years ago) and correlated with thrusting at Antler Peak and in the Sonoma Range. This sharply contradicts the younger (80 million years ago) dating by Merriam and Anderson.

These findings in Eureka County and nearby areas established a link with the work of Fergie, Si, and myself in the Sonoma Range quadrangle. By recognizing identical geologic relationships of thrust plates of oceanic rocks riding over shallow-water carbonate rocks of the same age span, we were able to trace the regional structural features throughout northern Nevada. Robert and I reported these new data at a GSA meeting in 1955.

Tom Nolan collaborated with others (1956) to summarize their work in the Eureka District. In this report they described the siliceous rocks of deep-water facies in ranges west of Eureka, noting that these rocks had come into the region on "thrust faults of very large displacement, which may have formed at various times from the late Paleozoic to early Tertiary." However, Tom once again reiterated his concept that a "late Paleozoic positive area" existed in late Paleozoic time in central Nevada. Tom considered this "positive area" to be a broad, simple upwarp of pre-Mississippian rock units extending across Nevada. We found, to the contrary, that this interpretation did not fit the facts. Rather, the rocks about fifteen miles west of Elko proved to be oceanic chert, shale, and volcanic rocks that had been thrust upon carbonate rocks of the same age span.

In the late 1950s, J. Fred Smith and Keith Ketner invited Tom to go with them to Ely, Nevada, where they were to present an informal paper on the geology of north-central Nevada. They outlined the major elements of the geologic framework, including the Antler Orogeny and related Roberts Mountains thrust. After the presentation, Tom is reported to have asked, "Is that the way it happened?" Both J. Fred and Keith said, "Yes, that is the way we see it." So Tom was finally persuaded that the Antler Orogeny was real.

Another report that helped clarify the late Paleozoic geology of the region was written by David Brew, one of Tom's assistants, who mapped part of the Diamond Range for a thesis project (1961). Under Tom's supervision, David clearly stated in his thesis that the Diamond Peak Formation in the Diamond Range contained much coarse clastic debris of Mississippian age derived from the Roberts Mountains thrust plate in the Antler Orogenic Belt on the west. David told me in January 1998 that he had freely discussed the Antler Orogeny and its implications with Tom.

In Prof. Paper 406, 1962, p. 27, Tom discussed the structure and stratigraphy of the Eureka area, especially the marked unconformity at the base of the Permian Carbon Ridge Formation. He also noted, about twenty-five miles to the northwest, an unconformity at the base of the correlative Garden Valley Formation, which rests on Ordovician chert and shale (of the Roberts Mountains allochthon). Tom went on to say (1962, p. 28), "Still farther west and northwest, Roberts and his coworkers (1958, p. 2850–54) have found extensive thrusting—assigned to the Antler Orogeny—which is of latest Devonian to Early Pennsylvanian age." Thus, by 1962 Tom recognized the late Paleozoic age of the Antler Orogeny, yet he still exempted the Eureka area from direct involvement in it. Larson and Riva in the mid-1960s, however, mapped quartzite and black shale in the Diamond Springs quadrangle about twenty miles north of Eureka that they assigned to the Vinini Formation of the Roberts Mountains allochthon.

Finally, in 1974 Tom came full circle. He wrote a paper entitled "Stratigraphic Evidence on the Age of the Roberts Mountains Thrust, Eureka and White Pine Counties, Nevada." He gave full credit to those of us who had worked on regional structural problems, such as the Antler Orogeny and Roberts Mountains thrust in Nevada. He implied, however, that orogeny was virtually continuous from late Devonian to late Tertiary time. Most geologists who have worked in Nevada, however, see major orogenies as being episodic.

My work in Eureka County provided a link between the geology of the Eureka area and that of Antler Peak. It was obvious to me that the Roberts Mountains thrust traveled east at least as far as Eureka, where it was overlapped by conglomerate of Mississippian age.

During those years I loved to roam far and wide throughout north-central Nevada. On one trip I joined William O. Vanderburg, a mining engineer with the U.S. Bureau of Mines, for a visit to Gold Acres, a mine owned by the New London Extension Company. Bill was engaged in reconnaissance mineral surveys of counties in Nevada. I thought his chore was impossible because he had to visit all the mining districts in the state, run down the mine operators, obtain production data from them, and make a brief, descriptive report on the host rocks of the ore from each mine, the mining methods, and any available metallurgical data.

Bill was a perceptive engineer, and he noted a curious fact about gold metallization in the region. At several mines, he had noticed that the gold was much too small to be recovered by panning. This was especially true at Gold Acres. Bill wanted to show me this unusual ore. My curiosity was piqued, so I went with him. Harry Bishop, Gold Acres's mine manager, showed us the ore along a great zone of shearing. The principal gold values, Bishop pointed out, were mainly in limestone lenses within the shear zone that separated upper-plate chert and shale from lower-plate limestone, identical to the situation that Merriam and Anderson had seen and mapped in the Roberts Mountains. For anyone who sees this contact zone in the field, the color contrast between the dark siliceous oceanic rocks and the underlying gray continental shelf carbonate rocks is striking!

The thrust contact at Gold Acres was not a narrow zone a few inches to a few feet thick as in the Roberts Mountains. Instead, it was a wide zone of shearing, a hundred or more feet thick, in which lenses of lower-plate limestone had been caught up in the thrust zone. These limestone lenses were the ore bodies!

FIG. 8.1. *View in Maggie Creek showing Roberts Mountains thrust plate (allochthon), dark rocks riding over gray limestone of the lower plate.*

These rocks told me that the gold deposits were part of the geologic framework! Sometime later, I accompanied Frank Maloney, a prospector sent by Governor Curley of Massachusetts, to evaluate gold mines in Nevada. Curley had heard of the potential for gold deposits near the Gold Acres and Getchell Gold Mines, and he wondered if another such deposit could be found. So Frank took me to the Copper King Mine a few miles northwest of Carlin.

When we entered Maggie Creek Canyon, I saw for the first time the stunning view of upper-plate black chert riding on iron-stained gray limestone on the east side of the canyon. At the Copper King Mine, I saw the clear-cut structural control of copper mineralization in sheared rocks in the upper thrust plate. This visit was a profound and moving experience for me as it confirmed that the thrust was of regional extent and might exert a regional control of mineralization. This had to be the same kind of relationship we had seen at Gold Acres—metallization controlled by thrust faulting.

Shortly afterward, Marion Fisher, a prospector and businessman from Battle Mountain, offered to show me his Bootstrap Mine about ten miles to the north of the Copper King. There I saw the same relationship of ore to structure that I'd seen at Gold Acres—a thrust fault

FIG. 8.2. *Author, with field glasses, leading group into Shoshone Range, stops at massive beds of Valmy Formation quartzite of the Roberts Mountains thrust plate, ca. 1956.*

separating upper-plate chert and shale from lower-plate limestone. Metallization was mostly in the sheared limestone, just below the thrust. The thrust plate apparently formed a cap rock, preventing ore-forming solutions from escaping into the rocks above, something like oil and gas being trapped under impervious rocks in a dome.

I had good reason to feel excited about the Gold Acres and Carlin-Bootstrap areas. My excitement grew as I visualized potentially rich ore bodies occurring along certain stratigraphic zones. I wondered if there were a recognizable pattern of the distribution of ore bodies in the region. I began to share my geologic picture with anyone who was interested—mining company geologists and engineers, USGS geologists, and university professors and students. Some understood the regional story right away and agreed that it made sense, but others did not believe that rocks could be thrust as far as I suggested—as much as one hundred miles from their source.

In mapping Eureka County, I was ably assisted by Robert Lehner and Mendell Bell. Our family lived on the Dean Ranch in a trailer. It was thirty-two feet long and had beds enough for all of us. Arleda and I slept in the bedroom, Mike and Steve in bunks in the middle, and Kim on a bed made up in the dinette. Bob and Mendell lived in a

Fig. 8.3. *Field camp in Crescent Valley at Dean Ranch (left to right): Michael, Mendell Bell, Steven, Arleda, Kim, and Bob Lehner, 1956.*

little log house nearby. The kids explored the nearby ghost towns of Cortez and Tenabo while we worked in the field. We traced the thrust from the Roberts Mountains through the Simpson Park Range, Gold Acres in the Shoshone Range, and into the Lynn district (Carlin area) in the Tuscarora Range.

Regional work also permitted us to trace the Battle Formation conglomerate east through the Cortez Range to Eureka. The conglomerate at Eureka was Mississippian age, late Mississippian age (we estimated) at Cortez, and Pennsylvanian age in Battle Mountain, thus prograding westward as it lapped onto the Antler allochthon. Ferguson, Muller, and Roberts (1952) described the Inskip Formation of Mississippian age in the East Range, and Preston ("Pres") Hotz and Ron Willden (1964) described a small pocket of Mississippian rocks in the Osgood Mountains. It is clear, therefore, that the Antler Orogenic Belt was active during Mississippian time here, within the western part of the belt.

Early on, Bob and I decided to point out to the world that the Roberts Mountains thrust was a major regional feature and significant in controlling ore deposits in the county. So we wrote an abstract and presented a paper in 1955 for the cordilleran section of GSA. In this

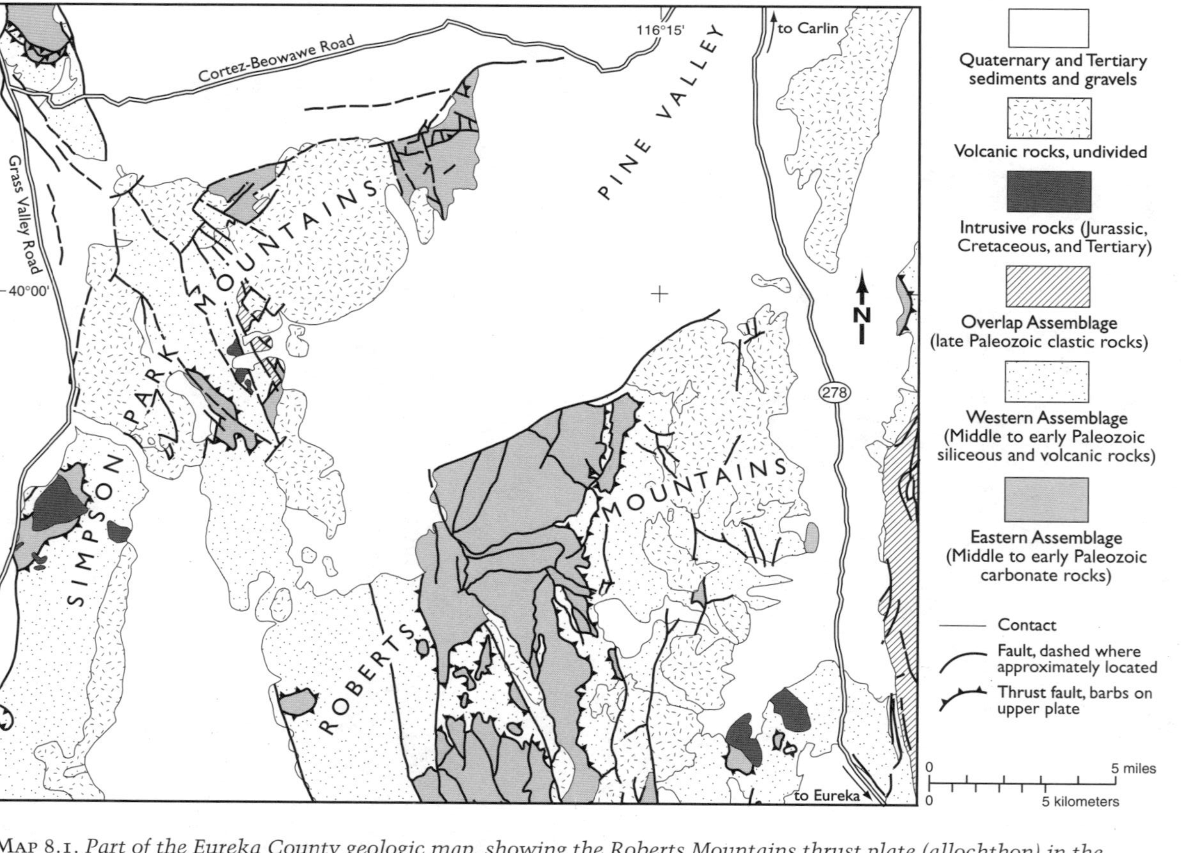

Map 8.1. *Part of the Eureka County geologic map, showing the Roberts Mountains thrust plate (allochthon) in the Roberts Mountains and adjacent ranges (coarse stiple) and underlying lower-plate carbonate rocks (fine stipple). Late Paleozoic rocks resting on the Roberts Mountains allochthon are shown with lined pattern.*

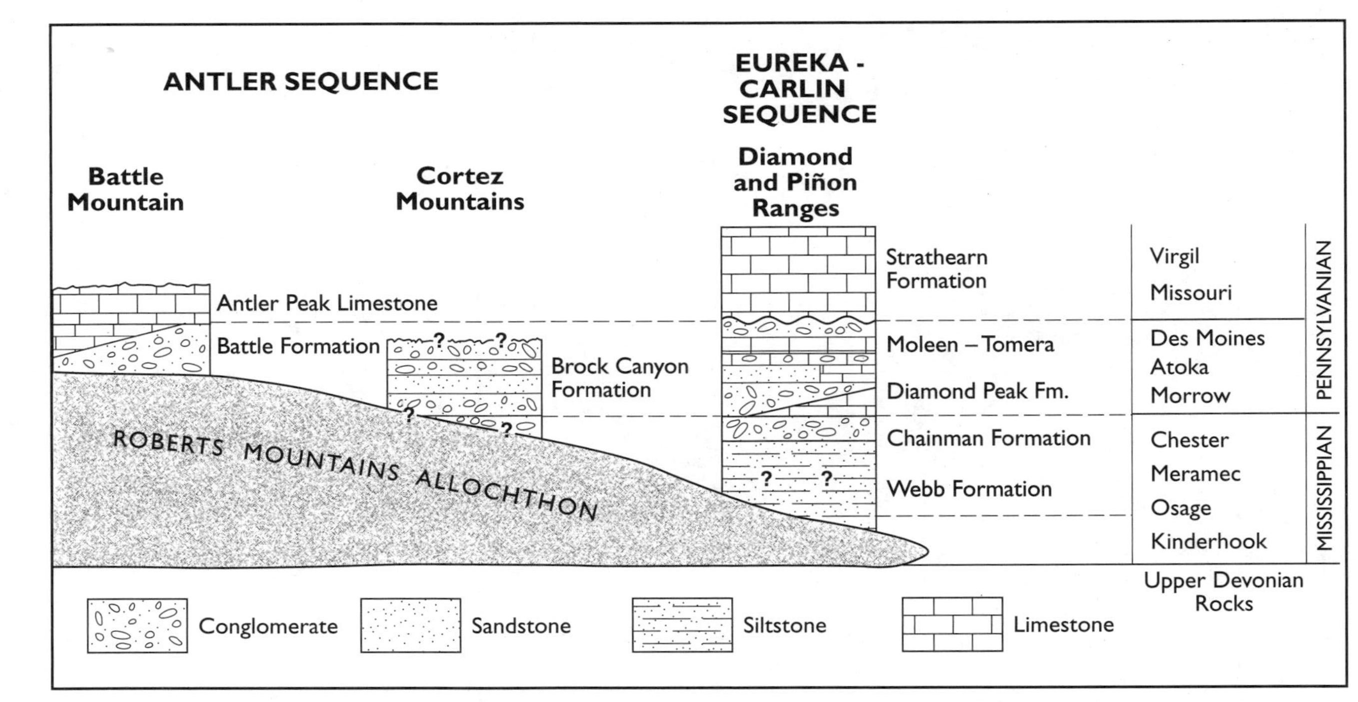

FIG. 8.4. *Westward onlapping from Eureka area to Antler Peak of late Paleozoic conglomerate facies onto the Roberts Mountains thrust plate (allochthon) during the Antler Orogeny. The onlapping thus shows that the orogeny developed from east to west across the region from Eureka in Mississippian time to Battle Mountain in Pennsylvanian time.*

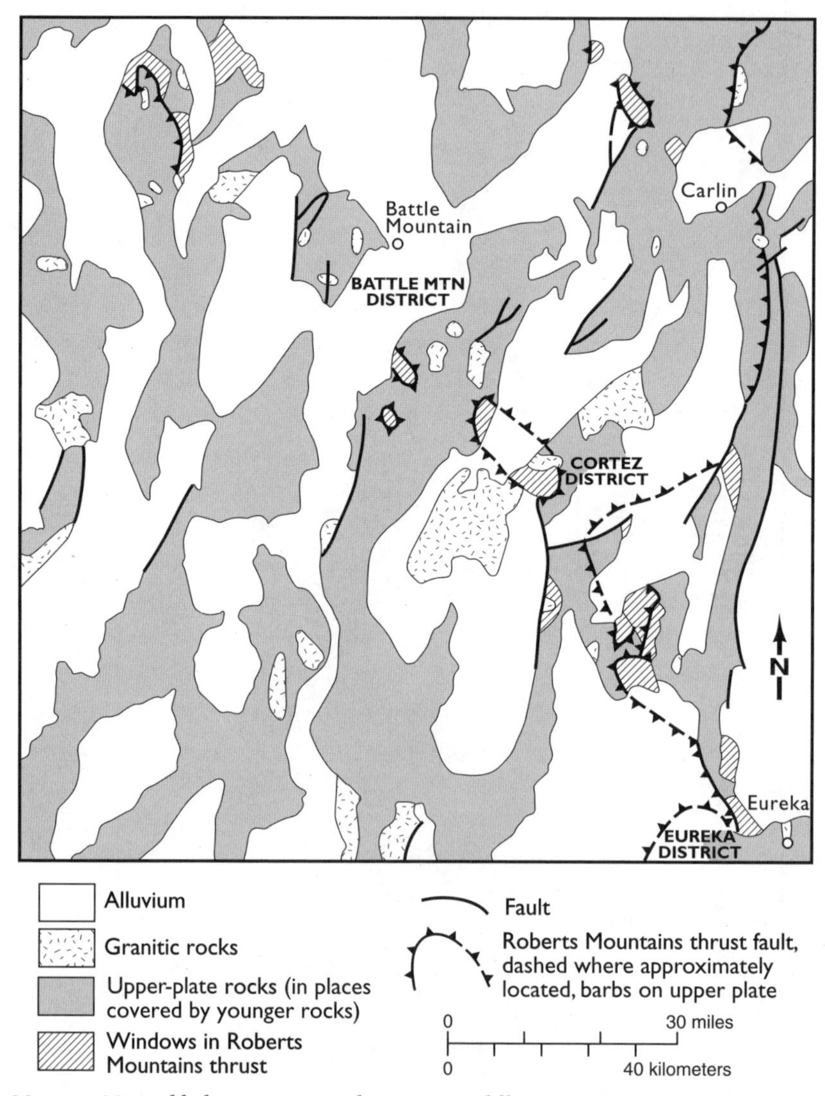

MAP 8.2. *Mineral belt map presented at 1955 Cordilleran Section meeting,* GSA, *Berkeley, California. The Lynn-Railroad Belt (now Carlin), upper right, and Battle Mountain\-Eureka Belt (now Battle Mountain), center.*

presentation, we included a crude map of the region (see map 8.2), plotting major mining districts, windows in the thrust, and some granitic bodies. Much to our surprise, we found that we had clearly shown northwesterly mineral belts in north-central Nevada, centering on the Battle Mountain–Eureka Belt, but also including the Lynn-Railroad Belt (now Carlin). These mineral belts gave our work a strong new focus.

We were aware that we had made a strong connection between regional geology and ore deposits. We also noted that windows in the Roberts Mountains thrust contained the principal mining districts in this region. Unfortunately, Bob Lehner left the project to join Bear Creek Mining Company in 1956, but another geologist, Kathleen M. Montgomery, helped me compile the Eureka County report.

A year later I gave further emphasis to the mineral belt concept in a paper at a Reno meeting of the AIME (American Institute of Mining Engineers). I called attention to the fact that mineral deposits were not randomly scattered around the state of Nevada but were mainly in structurally controlled zones of deformation. Thus, I told the audience, "you can concentrate prospecting along the mineral belts and save both time and money." The response was a deafening silence.

In 1958 Pres Hotz, Jim Gilluly, Henry Ferguson, and I assembled stratigraphic and structural data from many ranges in northern Nevada and published a landmark paper in the AAPG on the "Paleozoic Rocks of North-Central Nevada." We presented a coherent geologic summary of the region, clarifying the roles of the Roberts Mountains thrust and the related Antler Orogeny.

Later papers of ours (1964, 1965) shed additional light on regional geology, mineral belts, and metallization in Nevada. Although others could have written significant parts of this story, they failed to assemble the White Queen's "six impossible things before breakfast" and thus missed the greatest geologic plum of the twentieth century in Nevada.

I continued work in the Carlin area, establishing the concept that most ore deposits were in mineral belts, mainly in the carbonate rocks of the lower plate. I thought this was worth highlighting, so I wrote a brief report, which was published in the USGS yearbook in 1960, titled "Alinement of Mineral Deposits in North-Central Nevada." Included in the principal illustration were the Lynn-Railroad (now Carlin) and Battle Mountain–Eureka Belts. At that time, the Carlin Belt contained only two small gold mines (the Bootstrap and the No. 8) on its northern part. Thus, geologists could zero in on likely areas to prospect.

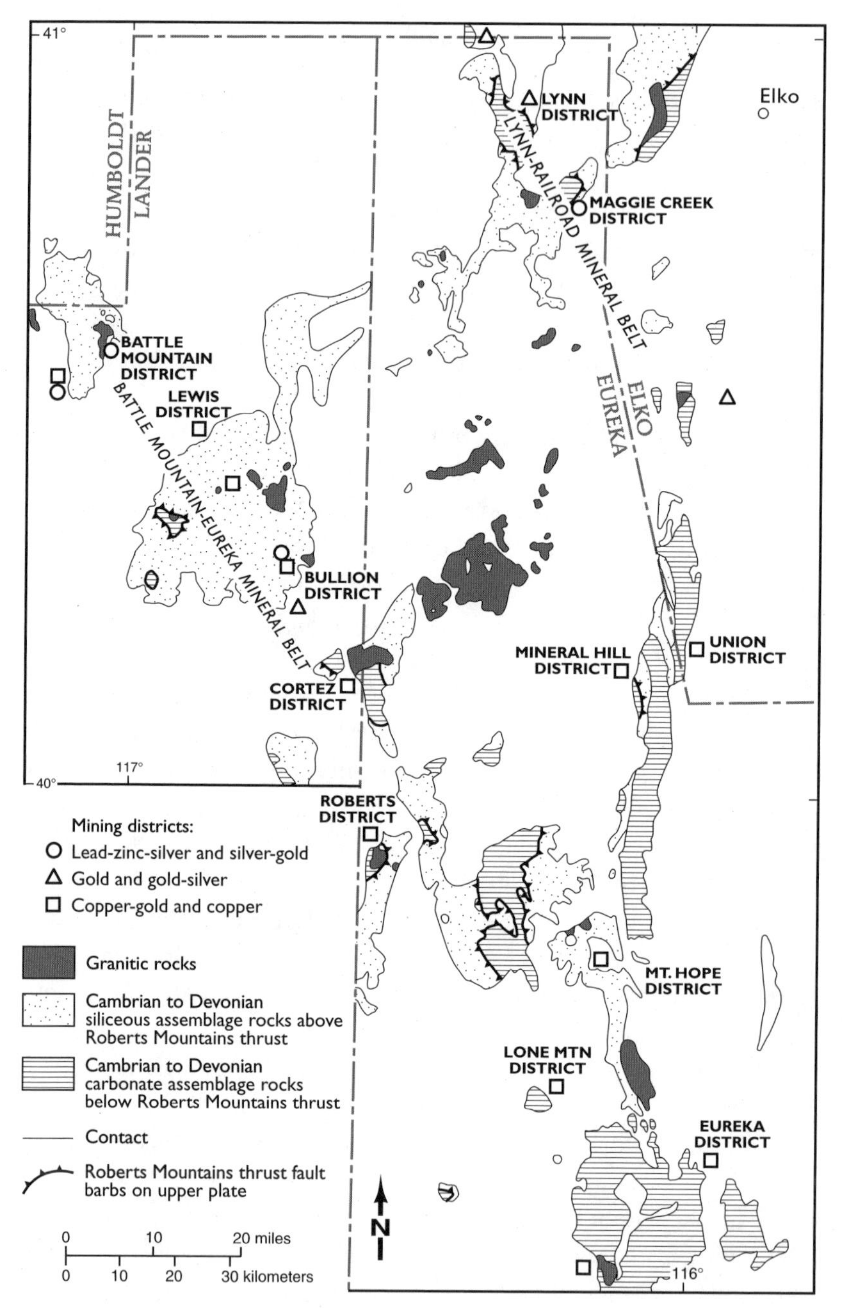

Map 8.3. *The Carlin and Battle Mountain Belts as presented at AIME meeting in 1957. Gold deposits in the Carlin Belt are Bootstrap (north) and No. 8 (now Genesis).*

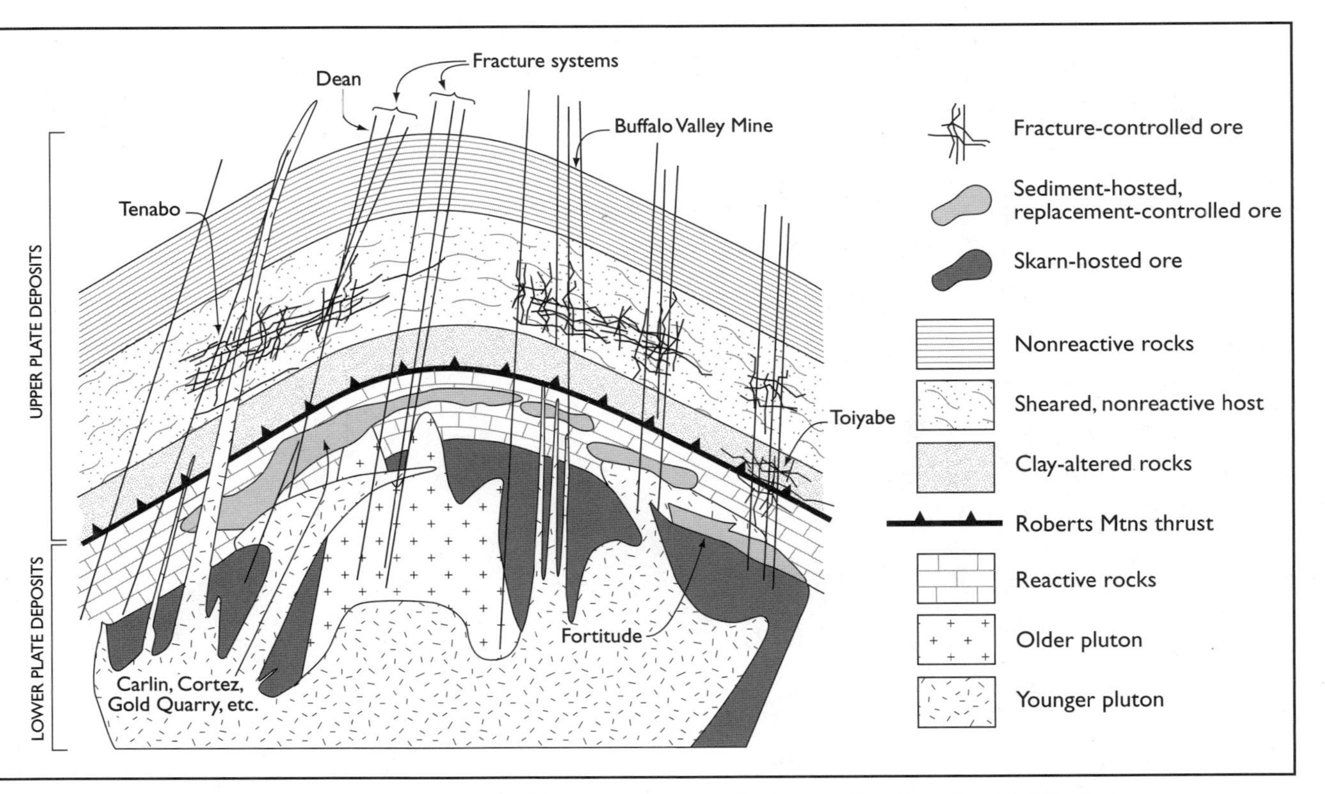

FIG. 8.5. *Rock types in Carlin area, cut by close-spaced fractures that controlled mineralization (after Raul Madrid).*

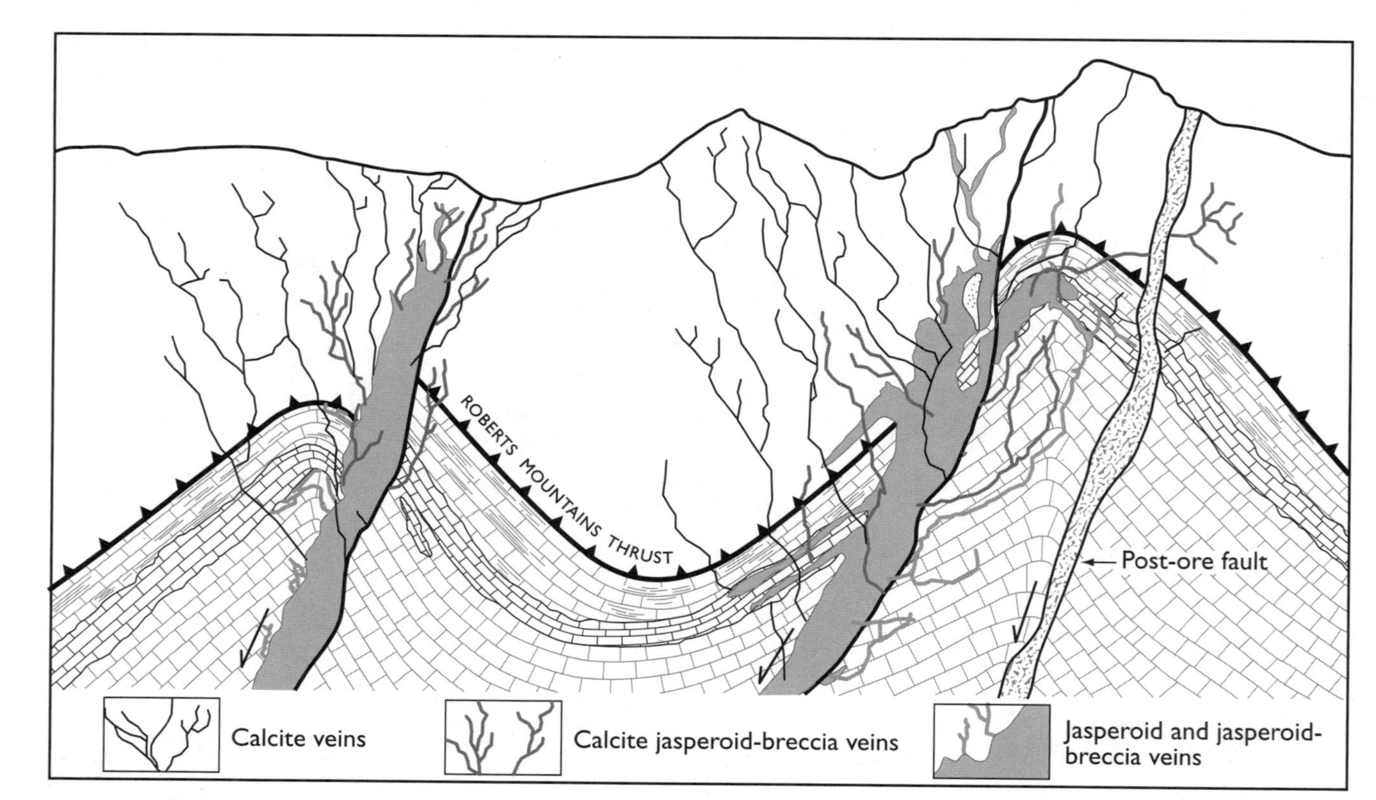

FIG. 8.6. *Folded carbonate rocks cut by mineralized fractures (black). The fractures permitted hydrothermal solutions to bring calcite and jasperoid (fine-grained silica) into mineralized zones, along with gold, silver, and other metals (after Raul Madrid).*

Later, in 1961, Lehner and others published a preliminary geologic map of Eureka County, and John Roen (1961) released a detailed map of the Carlin window in a master's thesis submitted to the University of California at Los Angeles; we used this map in compiling the Eureka County geologic map.

Raul Madrid and I later prepared illustrations showing the major features of ore deposits of Carlin type. Raul drew a composite cross-section of the principal stratigraphic and structural features and mineralization. (For a recent depiction of mineral belts, see map 8.4).

It seemed the last two pieces of the puzzle, the geologic controls of the gold deposits, were coming together. We had established that the Antler Orogeny and related Roberts Mountains thrust extended throughout central Nevada.

My list of "six impossible things" (begun in chapter 4) is now complete with the addition of the fifth and sixth elements: (5) thick limestone beds of Lower and Middle Paleozoic age, especially the Hanson Creek, Roberts Mountains, and Popovich Formations, and (6) the folds, faults, and fractures along mineral belts that permitted igneous rocks and associated gold-bearing hydrothermal solutions to rise into the upper crust. There they could react with carbonate-bearing rocks of Lower and Middle Paleozoic age, thus permitting the formation of immense gold deposits beneath the cap rock of the Roberts Mountains thrust plate.

To summarize the six ideas that integrate the geological framework with ore deposits in Eureka County, we have (1) impermeable oceanic cap rocks of the Roberts Mountains thrust plate, (2) gold deposits along and below the thrust, (3) the Antler sequence of carbonate-bearing rocks, (4) the Antler Orogeny, (5) the thick Lower and Middle Paleozoic carbonate sequence of north-central Nevada, and (6) the mineral belts, especially the Battle Mountain, Carlin, and Getchell mineral belts.

Other geologic units, including beds of Mesozoic and Tertiary ages, may also be mineralized in north-central Nevada. A diagram (see fig. 8.7) shows these units, including the Paleozoic units.

I later traced the Antler Orogenic Belt into central Idaho, to Copper Creek near Mackay where Betty Skipp and others (1979) confirmed

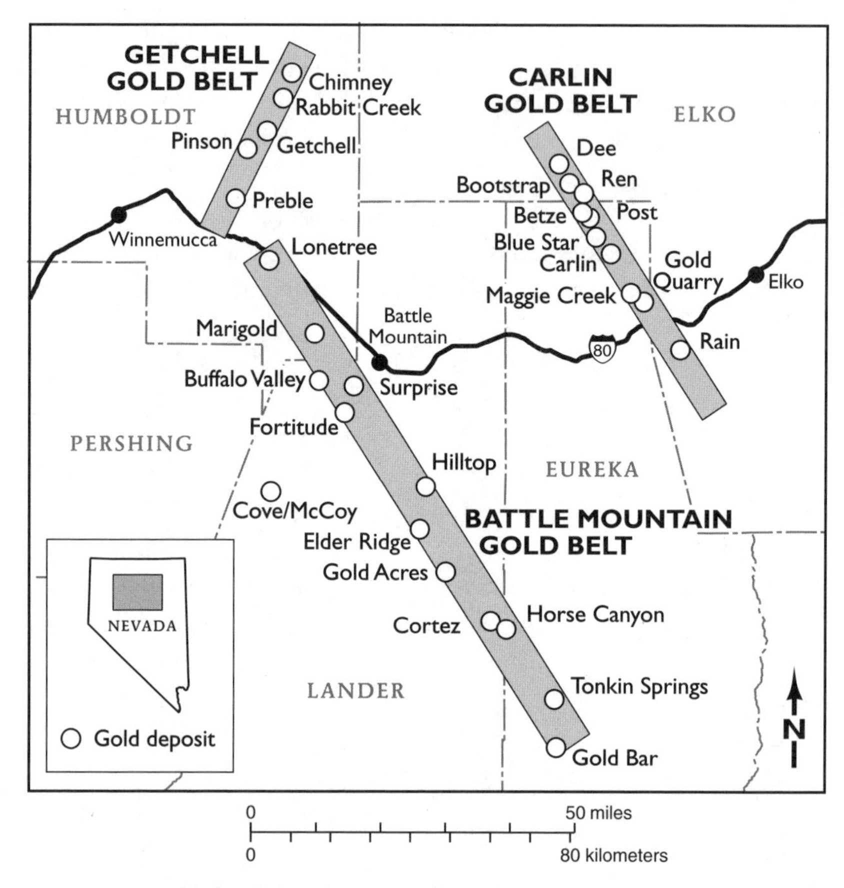

Map 8.4. *Mineral belts in Nevada, 1991.*

its presence, and into northeastern Washington where Charles Bennett (1936) described chert conglomerate of Devonian age. J. M. Mattison (1972) has dated intrusive activity in the Cascades of Washington at 265 million years ago, which coincides nicely with the Antler Orogeny. It seems likely to me that Carlin-type gold deposits might occur in Idaho, Washington, and even in British Columbia, Canada.

Elsewhere in the world, Carlin-type deposits could form with fewer steps, but here in the Carlin Trend all six steps are needed. Mackin's challenge to visualize six impossible things before breakfast proved useful! Following the White Queen's admonition to Alice

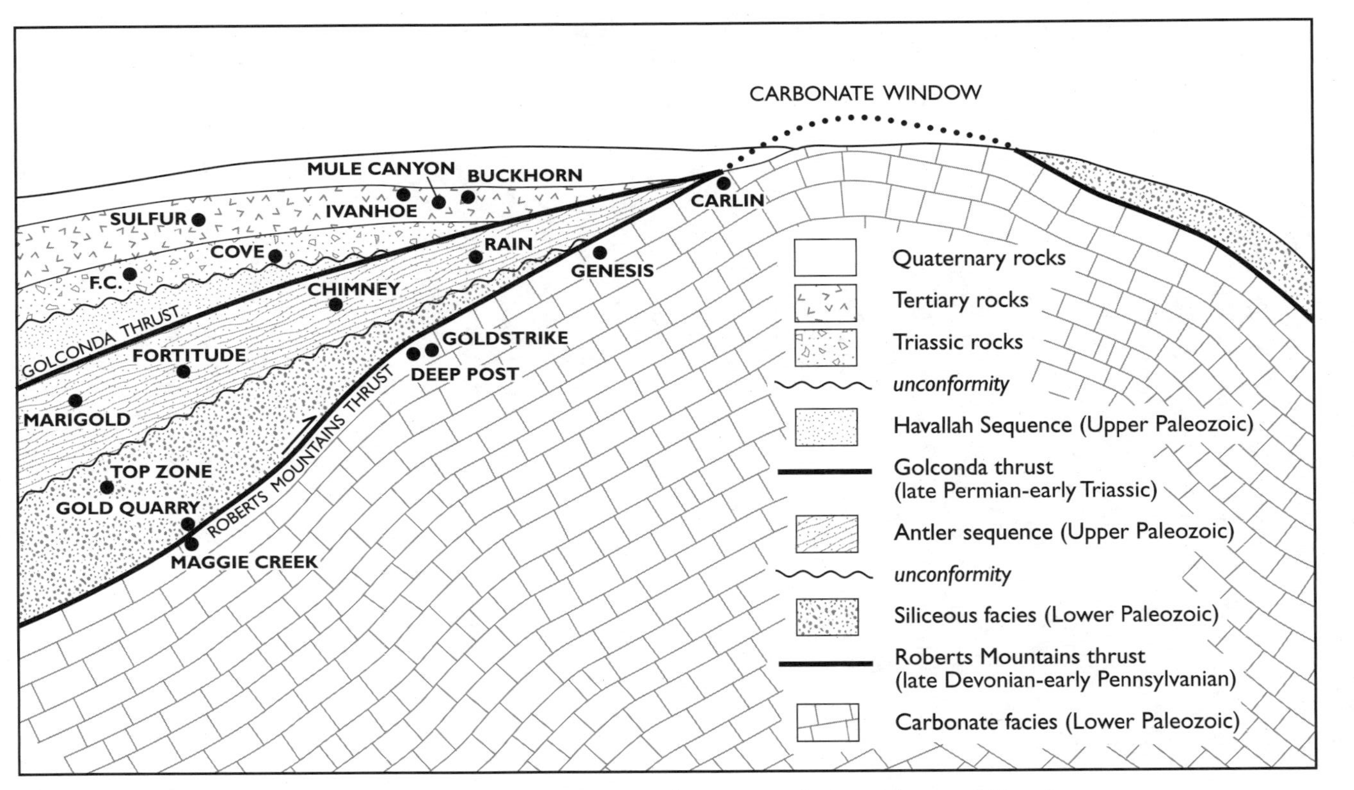

FIG. 8.7. *Possible sites for metallization in Paleozoic, Mesozoic, and Tertiary rocks in north-central Nevada.*

worked! The six geologic elements proved to be true. The gold is principally in the Pilot Limestone beneath the Chainman Shale.

In early summer of 1961, I was invited to present an informal paper on the regional geology of north-central Nevada before the Eastern Nevada Geological Society, a petroleum-oriented group, at a meeting in Ely, Nevada. I incidentally mentioned that the ore deposits, such as those in the Battle Mountain and Carlin Belts, fit into the central-Nevada geologic framework, and I showed a slide of the mineral belts that we had recognized. In my presentation I also mentioned the concept of localization of ore deposits on the "margins of windows." Although the presentation did not excite most of the petroleum seekers, one member of the audience seemed intrigued. He was John Livermore, an exploration geologist for Newmont Mining Company, and he had come from nearby Eureka to hear my talk. After the meeting, John was waiting to introduce himself. We sat down to talk, and John mentioned that Fred Searles, chairman of Newmont Mining Company, and Robert Fulton, Newmont's chief geologist, had sent him to Nevada to find a large gold deposit. John asked me to discuss the position of ore deposits in my structural framework in more detail. So I outlined the geologic elements that might interest him, especially the geologic picture at Gold Acres and the Bootstrap, where the gold was clearly tied to regional thrusting and was localized in mineral belts.

John and an associate, Alan Coope, had read our 1957, 1958, and 1960 papers and used our geologic framework in exploration. After John and Alan had been in the area that summer, John invited me to come out and visit them in the field in early October to help identify rock units of the Carlin area. We went to the summit of Popovich Hill, which was composed of Devonian limestone. I suggested that the underlying Roberts Mountains Formation would be a good exploration target as it was the host rock at the nearby Bootstrap Mine. John and Alan explored the ground just south of Popovich Hill, and discovered a zone 80 feet long in a bulldozer pit that averaged 0.20 ounces of gold per ton. The rest is history. John Livermore and Alan Coope then drilled a hole nearby and cut a mineralized zone that contained more than an ounce of gold to the ton. John and Alan are thus credited with the Carlin discovery in late 1961 (Coope, 1991).

FIG. 8.8. *Aerial view of Carlin No. 1 Mine, showing the mill (right) and pit (center).*

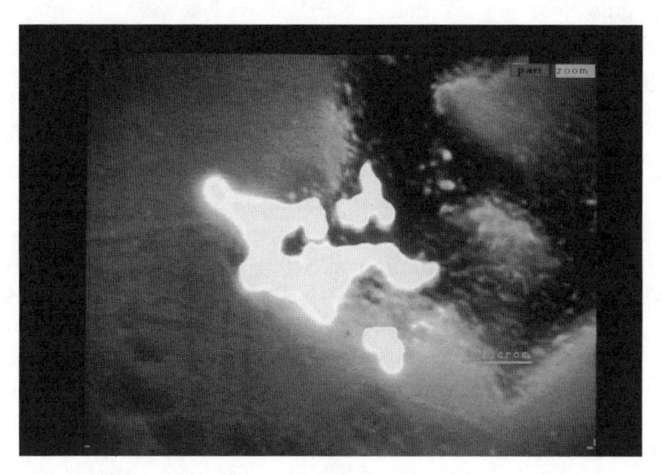

FIG. 8.9. *Electron photomicrograph of gold in Carlin Mine ore. The gold is the light spot in the center surrounded by grayish clay minerals. Courtesy of Newmont Mining Company. A one-micron scale is shown in upper right.*

After the discovery, John and Alan carried on an aggressive drilling program. The mineralized area became the Carlin No. 1 Mine and yielded more than 11 million tons of ore, containing more than 3.5 million ounces of gold, averaging about 0.32 ounce gold per ton. This ore warranted building a mill, which Newmont completed in 1965.

The submicroscopic nature of the gold at first was no problem, but carbonaceous ore in deeper levels had to be oxidized before the gold could be recovered.

Tom Nolan, director of the USGS; Roy Hardy, undersecretary of the Department of the Interior; and I were invited to the dedication of the mill. *Newsweek* wrote about the dedication as follows:

Gold: "A Miner's Dream"

> When it was published in 1958, a slim, yellow paperback called "Paleozoic Rocks of North-Central Nevada" did not seem an especially promising candidate for the best-seller lists. It turned out to be a gold mine. Not for the author, Ralph Roberts, 54, a veteran U.S. Government geologist who had spent years slogging across the reddish, sage-flecked hills of northern Nevada. But the sharp eyed geologists of New York's Newmont Mining Company took a leaf out of Roberts' book and started an intensive exploration of the area. Using Roberts' leads, they struck gold. So much of it that in Carlin, Nev. last week Newmont dedicated the biggest gold mine to open in the U.S. in this century. Plato Malozemoff, president of Carlin Gold Mining Company, the Newmont subsidiary that will operate the open-pit mine, estimated that it holds 3,520,000 ounces of gold—or enough to keep running at full capacity for at least fifteen years. He predicted that it would yield $7.8 million in gold a year (about one seventh of total U.S. output). "It's a miner's dream," he exclaimed. . . . How does Roberts, who has now moved on to Utah to map rock formations at $18,580 a year for the Geological Survey, feel about missing out on Carlin's bonanza? "It's the role of government [geologists] . . . to point out possible target areas and leave it up to private industry to explore them," he says calmly. "I just get pleasure out of doing my work." (June 7, 1965, pp. 71–72)

Of course, the discovery led to renewed interest in gold exploration in north-central Nevada. The painstaking fieldwork by Bob Lehner, Mendell Bell, and myself in Eureka County during the preceding years had paid off handsomely.

In 1967, Don Hausen, of Newmont Mining, and I were invited by the chairman of the department of geology to the University of St. Andrews, Scotland's oldest university, to present papers on the Carlin discovery. My coauthors, Keith Ketner and Arthur Radtke, and I defined, for the first time, Carlin-type deposits as a new class of disseminated gold deposit, characterized by gold-mercury-arsenic-antimony mineral assemblages: the silver and base metal contents are low.

The trip was memorable for me in another way, as well. Through the courtesy of a Professor Campbell in the geology department we were invited to play golf on the venerable Old Course at St. Andrews, the birthplace of golf. Despite the gorse bushes in the rough and infamous pot-bunkers, I shot a score of eighty-seven on my second round, not too bad for a twelve-handicap American player on a Scottish links course.

In 1968 Don Hausen and Paul Kerr published a paper on the occurrence of fine gold at Carlin. In 1971, Arthur Radtke, Robert Coats, and I wrote a summary paper on gold deposits in north-central Nevada and southwestern Idaho in which we further defined Carlin-type gold deposits as low temperature replacement deposits characterized by a trace element suite of arsenic, mercury, antimony, and thallium minerals. In 1997, with much more data available, I further defined Carlin-type deposits as characterized by: (1) a low temperature mineral assemblage, mainly pyrite, arsenical pyrite, gold, stibnite 'orpiment, and realgar, and minor cinnabar, which formed mostly between 280 and 150 degrees Celsius; (2) anomalous traces of arsenic, antimony, mercury, and thallium, and low silver content; and (3) largely disseminated, finely particulate, "invisible" gold.

Some Carlin-type ore bodies are world-class deposits (i.e., containing more than 5 million ounces of gold). Gold grades range from about 0.01 ounces to more than an ounce to the ton. The grade of ore mined is dependent on the price of gold.

Operation of Newmont Mining Company Following 1965

As time went on, Newmont Mining Company acquired other properties in the Carlin Belt. These included the Genesis (No. 8), Bootstrap, Rain, Post, Lantern, North Star, and Deep Star Mines and part of the Bullion-Monarch. In the mid-1980s Newmont Mining Company spun off the gold mines into the Newmont Gold Company.

In a letter in July 2000, Odin Christensen summarized the recent history of Newmont Gold Company, which I have paraphrased as follows:

> Newmont Gold acquired the gold properties of Santa Fe Railroad, including Twin Creek(s) in the Osgood Mountains, and Lone Tree and Stone House properties in Battle Mountain. In June 2000 Newmont Gold acquired Battle Mountain Gold Company, which had properties in the Battle Mountain area and near Chesaw in north-central Washington State. Christensen also pointed out that Newmont Gold had pioneered in "Metallurgical innovation-development of technology to process low grade and refractory ores; . . . Environment-conducting all activities to high standards of environmental protection."

Newmont Gold is also carrying out aggressive exploration and mining activities in Uzbekistan, Mexico, Indonesia, and South America.

The Cyanide Process for Gold Recovery

The fine grain of the gold in Carlin-type deposits requires metallurgists to use chemical processes for its efficient recovery. The gold-bearing ore is mined, crushed, and ground fine in a mill and then leached with cyanide solutions, commonly sodium cyanide (NaCN). The gold can be recovered from the cyanide solution by adding finely powdered zinc to the gold solution, which precipitates the gold and zinc together. This mixture can be heated, then the zinc sublimed off, leaving the gold behind to be melted and cast in bars. The bars contain silver and other metals and must be further refined before the gold can enter the world market. Many environmentalists are wary of the long-term consequences of the cyanide method of gold recovery, but fortunately, the cyanide solution spontaneously breaks down after a short time into harmless carbon and nitrogen. Ponds containing active cyanide solution must be covered to prevent access by birds and some animals.

Next—the Bingham Copper-Gold Project

When the Eureka County project work was well in hand, I joined Edwin Tooker in Utah to begin work on the Bingham copper-gold project in the Oquirrh Mountains. This project involved another kind of geology but promised an interesting challenge.

Chapter 9

The Oquirrhs

BINGHAM COPPER-GOLD PROJECT, 1956–1971

> The basic building block in field geology is the geologic framework. Until this framework is established, the complex geology of the Oquirrh's could not be unraveled.
>
> —RJR

> . . . let us trace the pattern, however disconnected . . . [it seems]. . . .
>
> —VIRGINIA WOOLF

In the center of the Oquirrh Mountains, just twenty miles southwest of Salt Lake City, is a huge open pit that stretches more than a mile from side to side and plunges more than six hundred feet from rim to floor. The mine has yielded impressive amounts of base and precious metals, yet no one seems able to answer the question "Why is the Bingham copper-gold ore body here?" The USGS was determined to find the answer. Ed Tooker and I accepted the challenge.

Edwin W. Tooker earned his Ph.D. from the University of Illinois, specializing in clay minerals. In the Bingham project Ed focused on stratigraphy and structure, an ideal approach. So we set to work, making careful stratigraphic studies to provide a firm basis for our structural interpretations of the Oquirrh Mountains, the Bingham District, and the ore deposits.

Ed and I set up camp in Tooele (locally pronounced Twilla), directly west of Bingham Canyon. Tooele was a perfect place to start our work;

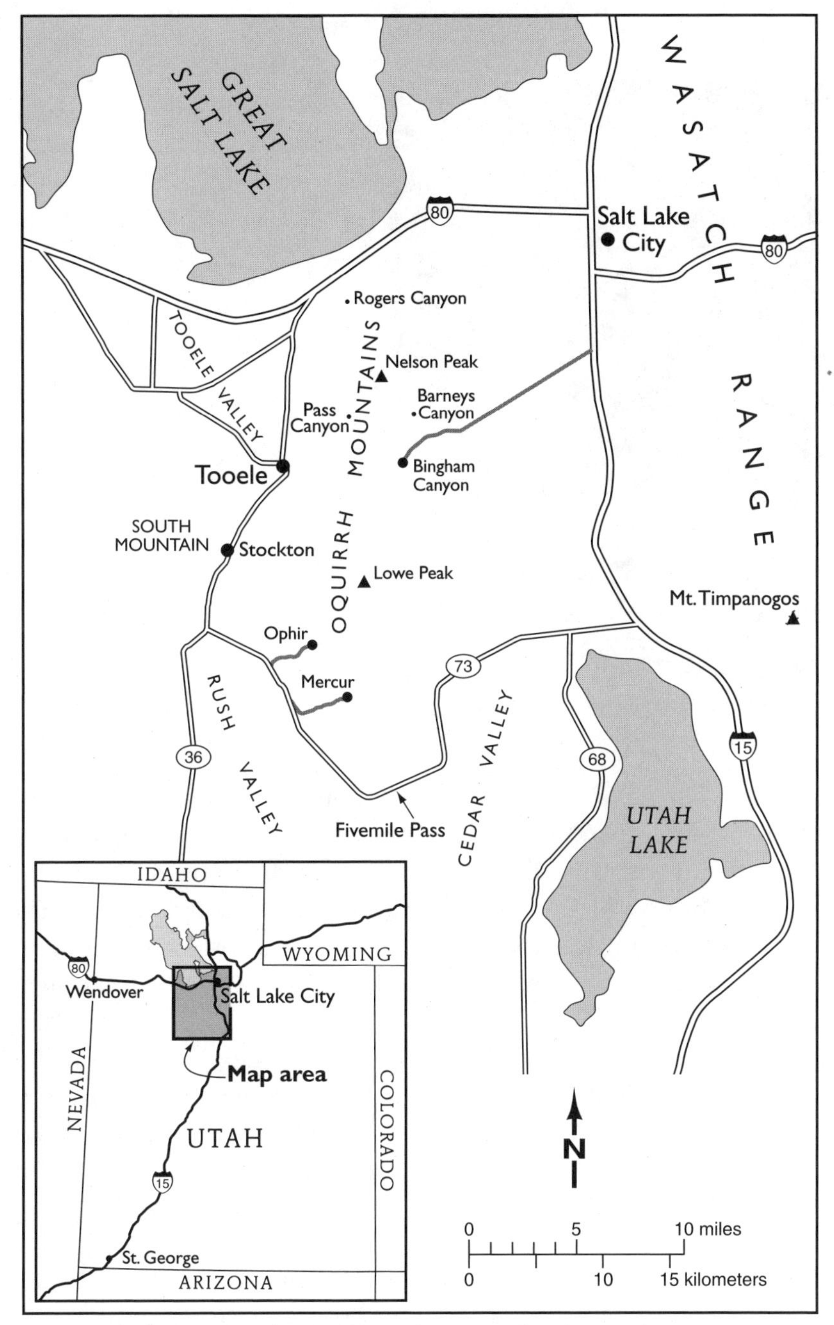

MAP 9.1. *Central Utah with Salt Lake City, Great Salt Lake, Bingham Canyon, Tooele, Stockton (mining district), Ophir, and Mt. Timpanogos (center).*

FIG. 9.1. *Edwin W. Tooker.*

it was the site for a U.S. Army ammunition dump and home to many of the miners employed by Kennecott Copper Company, owner of the Bingham Canyon Mine. It lay tucked away at the western edge of the northern Oquirrh Mountains, the Great Salt Lake to the north. Today, Tooele is a bustling town about twelve miles south of Interstate 80.

We towed trailers to our camp in Tooele, and we and our families lived in them. Our boys fell right into the life of the town. Mike coached a Little League baseball team; Steve played catcher on the team and quickly endeared himself to his teammates by hitting a home run in his first game. Kim played golf at the local course. Ed and I hiked with them up onto Nelson Peak; from there we had a wonderful view of Salt Lake Valley to the east and Tooele Valley to the west.

The principal objective of this project, as we stated in our report, was "to map the district and adjacent areas to determine the structural and stratigraphic controls of the intrusive rocks and related ore deposits." This seemed straightforward enough, and we tackled the job with vigor. We quickly submerged ourselves in the project. Our system was simple: we measured stratigraphic sections carefully, then traced beds along the slopes, mapping details as we walked. When we completed our stratigraphic studies, we started a mapping program, knowing that we could trace the units with confidence.

Early on, we contacted the Kennecott Copper Company geologists at the Bingham Canyon Mine and asked if we could meet with them to discuss our project and to consider joint projects. We thought they might accept our proposal more readily if we showed them a little hospitality, so we invited them to a picnic at our trailer park. They accepted and we all went into action—Polly and Ed Tooker, Arleda and I, and Mabel and Max Crittenden from the USGS project in the Wasatch Range. Anytime we all got together, we were guaranteed plenty of laughs. "The girls" had a reputation for putting tremendous energy, enthusiasm, and drama into just about anything they did, and as a result, our get-togethers were never ordinary. We put together a great meal—simple but wonderful hors d'oeuvres, a huge salad, roasting ears of fresh corn, and a hot, spicy tamale pie. The Kennecott geologists seemed quite impressed and easily joined in the camaraderie. Plates were filled and quickly emptied, and they didn't hesitate to help us finish off any would-be leftovers.

Polly, Arleda, and Mabel then disappeared into one of the trailers to put the final touches on what they had promised would be a showstopper of a dessert. They didn't emerge from the trailer right away, and I was just about to check on them when they suddenly appeared at the door, trying to appear composed. They were carrying a huge, beautiful watermelon, which had been carved and hollowed out and then filled with balls of watermelon, cantaloupe, and ice cream. They served it with great aplomb, and only later, after the Kennecott geologists had left, did we learn the reason for the delay. Their creation had just received its final touches when it slid off its tray and crashed to the floor. Miraculously, the melon boat had not broken, but ice cream and melon balls had gone rolling all over the floor. Without missing a beat, they had chased and scooped up the ice cream and melon balls, and gone on with the show. As they reenacted their story with great drama, we couldn't help laughing just as hard as they were.

Although our party was successful, our hopes for a cooperative program were not realized. The Kennecott geologists politely explained that they had already written reports on various areas and geologic problems we were interested in, and they feared the consequences to them if we came up with different interpretations. If our data were different from theirs, they would be forced to change their reports, and that could mean dismissal; at the least, their positions with the company would be undermined. We were disappointed, yet we had to accept their position. Later we learned that rumors had reached them of "that wild man Roberts," who found thrust faults wherever he worked, and they weren't ready to fully trust me. So, throughout our work on the Oquirrh project, we worked in close proximity but never collaborated.

Although the Bingham area had been mined for copper, gold, lead, zinc, and silver for more than one hundred years, the detailed geology of the Oquirrh Mountains had never been mapped completely. Ed and I started with a clean slate, building up our geologic picture as I had done in north-central Nevada, treading the ridges and valleys and walking out the folded and faulted fossiliferous limestone beds of the Oquirrh Range and describing the rocks as we went.

Of course, others had preceded us, especially Armand Eardley (1934), who identified the Nebo thrust in the Wasatch Mountains, and A. A. Baker (1947), who recognized its northern continuation, the Charleston thrust. They found that the Pennsylvanian and Permian rocks in parts of the Wasatch were extremely thick, measuring 26,000 feet (7,925 meters) thick near Mt. Timpanogos, about thirty-five miles southeast of Salt Lake City. Their work set the stage for ours.

During the 1950s and 1960s, Morris, Lovering, and Crittenden of the USGS, aided by faculty and students from Utah universities, were engaged in stratigraphic and structural studies of Paleozoic rocks in central Utah. Ed and I joined them in 1955, working specifically on the Oquirrh Range. We found that the mountains consisted of multiple thrust sheets, or nappes. Ed ultimately defined five major nappes that had come into the range sequentially from different directions. These nappes were composed of significantly diverse sequences of rocks derived from different parts of the Oquirrh basin. (The northern

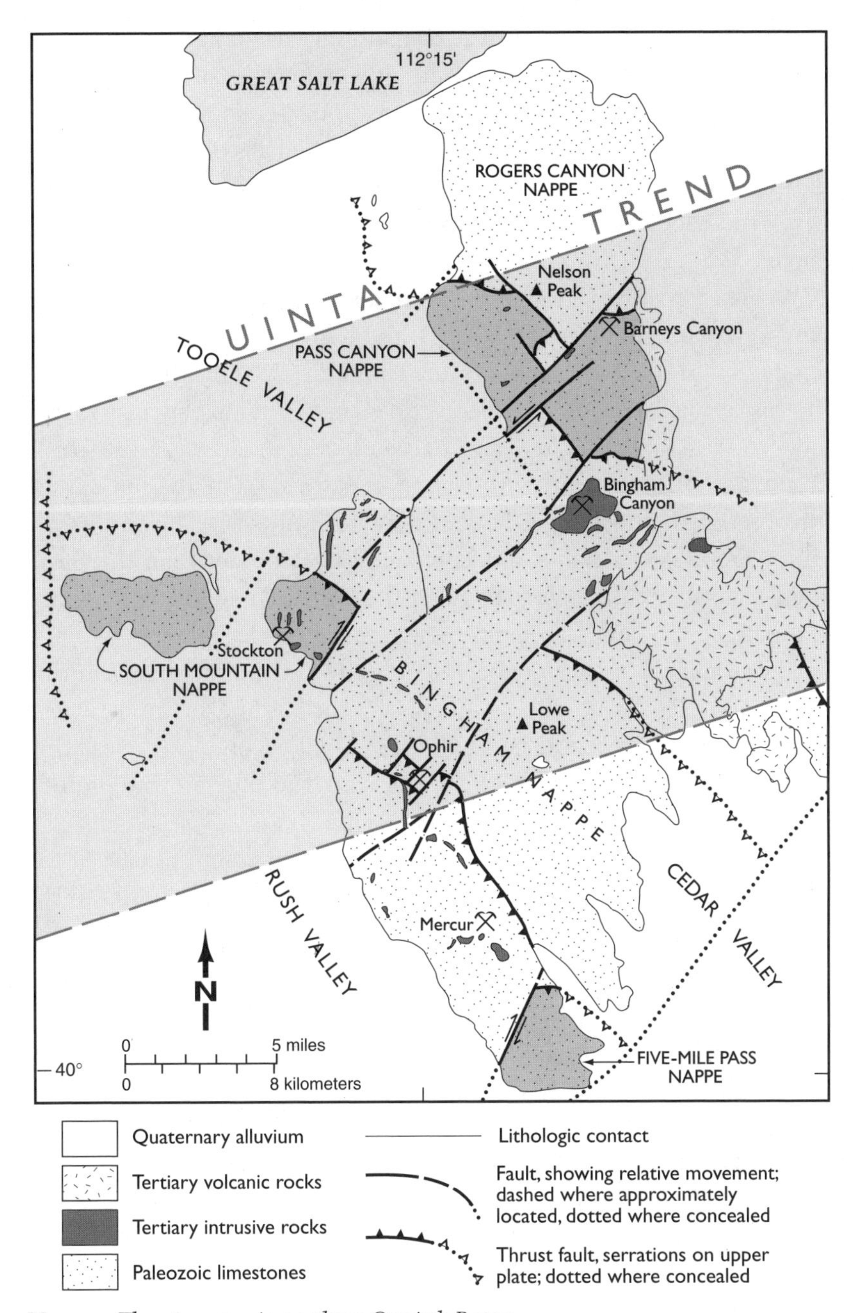

MAP 9.2. *Thrust nappes in northern Oquirrh Range.*

part of the basin is referred to as the Sublette basin.) Our task was to determine the detailed stratigraphy and structures of the nappes as well as how the ore deposits fit into this puzzle. Ed and I intuitively knew that the ore deposits in the range implied a geologic framework that had hitherto escaped previous workers, and we intended to discover the secrets of the Bingham and other mining districts in the range.

We read reports on the regional geology of the area, such as Boutwell's Bingham report (1905) and Gilluly's Stockton-Fairfield report (1932). Neither report offered much help. Gilluly noted that the eastern boundary of the Basin and Range Province lay just east of downtown Salt Lake City, and he went on to state that the Uinta Trend, a prominent east-west feature in eastern Utah, did not extend into the Basin and Range. Ed and I considered this to be odd as the Uinta Trend was a major orogenic trend much older than the Basin and Range structures. We found that the Uinta Trend had been a zone of great activity since Precambrian time. In the late Precambrian, it had been an axis of subsidence, forming the Uinta Trough, but during subsequent Paleozoic time, it had become a zone of intermittent uplift. Then, in Mesozoic time, the trend had become a buttress that influenced the movement of nappes into the site of the Oquirrh Mountains.

Ed and I considered the Uinta Trend to be an essential part of the geologic framework that controlled the position of ore deposits in the Oquirrh Range. In our reports, we interpreted aeromagnetic maps that showed the Uinta Trend clearly extending into the Oquirrh Range. Zietz, Roberts, and others (1969) confirmed this projection and showed that the Uinta Trend could be traced into eastern Nevada.

We also found that the Bingham porphyry bodies were emplaced at the intersection of an overturned fold in the Bingham sequence near the westward projection of the Uinta Trend. This structural intersection therefore became the site for intrusion of the stocks, dikes, and their related ore bodies, not only at Bingham, but also at Mercur, Ophir, Stockton, and the Barneys Canyon–Melco deposits.

We built up a strong stratigraphic and structural base for the control of the Bingham district ore bodies, which includes one of the great copper-gold ore bodies in the United States. Although Bingham originally was a copper mine, the production of gold has been and continues to be notable because of the large volume of ore treated annually. Bingham is a classic zoned, copper-molybdenum porphyry-hosted ore deposit with important associated lead-zinc and silver deposits. A central copper-molybdenum core grades out into a copper-gold zone and is partly surrounded by the peripheral lead-zinc-silver zones.

Production, 1863–1992

Copper	14,700,000	tons
Molybdenum	>387,000	tons
Gold	18,481,000	ounces
Silver	252,903,000	ounces
Lead	2,400,000	tons
Zinc	1,000,000	tons

The Bingham porphyry and associated Last Chance stock are hosted in the Butterfield or Bingham Mine Formation of the Oquirrh Group of Pennsylvanian age (Tooker and Roberts, 1970). These beds were deposited in the Oquirrh basin in western Utah and were thrust about seventy to eighty miles east and southeast into their present position in early Cretaceous time during the Sevier Orogeny (about eighty million years ago) (Roberts and others, 1965).

Ed (1996) noted significant differences in the stratigraphy in various parts of the Oquirrh Range and divided the rocks into five separate nappes: the Pass Canyon, Bingham, Rogers Canyon, South Mountain, and Five-Mile Pass. These nappes moved along different paths from the western basin hinterland onto an eastern foreland.

Ed has described the geology of the Oquirrh Mountains in detail. I will paraphrase his account as follows: The Bingham nappes converged on the Uinta Trend, producing folds and imbricate thrusts. The nappes contain distinctive sedimentary rock sequences, roughly equivalent in age.

In order of their arrival in the Oquirrh Mountains, they are:

1. Pass Canyon nappe, north-central part, was the first to arrive, moving eastward. The sole, or basal thrust on which it moved, is not exposed in the range. The Rogers Canyon nappe came from the north and Bingham nappe from the south. The basal thrust may be part of the Nebo-Charleston nappe identified in the Wasatch Mountains by Baker and others (1949).
2. Bingham nappe, in the southern half of the range, is the largest thrust plate. It moved eastward on the Midas thrust against the Uinta Trend and Pass Canyon nappe. The sole thrust was overridden by the Rogers Canyon nappe. The Bingham nappe contains a 7,989 meter–thick (over 26,000 feet) Paleozoic section, ranging from the Cambrian Tintic Quartzite to the Pennsylvanian Bingham Mine Formation. Normal faults developed during the Cretaceous cut Sevier thrusts and permitted the introduction of magma and hydrothermal solutions. Some of these faults were reactivated during Basin- and Range-tectonism.

   The Bingham base- and precious-metal ores are in intrusive rocks, veins, and adjoining limestones near the Midas thrust. Disseminated gold in the Mercur mining district is in Upper Mississippian carbonate rocks of the imbricate Manning thrust. The vein and replacement deposits in the Ophir mining district are localized mainly along normal faults and in Lower Paleozoic carbonate rocks.
3. The Rogers Canyon.
4. South Mountain.
5. Five-Mile Pass nappes have not yet yielded significant ore deposits.

Once the stratigraphy and structure of the Oquirrh Mountains had been worked out, the setting of the ore deposits could be tackled. The Barneys Canyon gold deposit, Pass Canyon nappe, five miles north of the Bingham pit, is a disseminated gold deposit with minor silver. It is in the Grandeur Member of the Park City Formation and upper Dry Fork unit on the north flank of the Uinta Trend. The deposit is controlled by a northeasterly trending fault in Barneys Canyon (Presnell and Perry, 1995). No granitic body has been found nearby, but an altered porphyritic dike occurs west of the deposit.

R. D. Presnell and W. T. Perry (1995), on the basis of potassium-argon radio-isotope dates of the clay mineral illite in the deposit, assigned a Jurassic date (about 140 million years) to the mineralization. Ed and I favor a more recent date, Upper Eocene, because the mineral assemblage of the Barneys Canyon deposit is similar to that of the Bingham porphyry, for which we have dates of 40 million years.

Ed (1996) proposed a geologic model "for the Bingham district that included Late Cretaceous thrust faults of the Sevier Orogeny, which brought five distinct nappes eastward onto the Uinta Trend. Mining districts occur in three nappes that overlie the Uinta Trend. The structures that localized the districts include arc-stressed, asymetric to locally overturned, thinned, leading edges of basal and/or imbricate thrust faults. The base- and precious-metal ore deposits are commonly spatially co-located with intrusive rocks." The South Mountain nappe contains the Stockton, Ophir, and Mercur mining districts. The Stockton district has yielded ores of both base and precious metals from the Stockton thrust, which overlaps the Bingham nappe (Tooker and Roberts, 1992). The Ophir district is in the Ophir anticline, which is composed of Lower to Middle Paleozoic rocks that are highly faulted and contain ore deposits with base and precious metals. It was productive during the early 1900s. The Mercur district, six kilometers south of Ophir, yielded significant gold from the 1970s through the mid-1990s. The district is on the upper plate of the Manning thrust; intersecting northwest and northeast—striking steep—dipping normal faults in carbonate rocks of Mississippian age.

Although Ed (1996) did not specifically target the problem of melting in the subcrust to produce the Bingham, Last Chance, and other stocks and their related ore bodies in the Oquirrh Mountains, one can speculate about the origin of the igneous rocks and ore bodies. A closer look at the Bingham and associated mining districts may offer some clues.

First, consider that these mines are near the intersection of the

Uinta Trend and the margin of the Basin and Range Province. This would be a zone of fundamental weakness. Now, imagine several thick nappes from the west and northwest piled onto this zone. The weight of the Bingham nappe alone, totaling nearly eight thousand meters—roughly twenty-six thousand feet to which we can add the cover of Mesozoic, and early Tertiary rocks, an additional fifteen thousand feet, represents an enormous load. This load would depress the rocks along the Uinta Trend in late Mesozoic and early Tertiary time some five to seven miles deeper into the crust.

From this point on I am speculating, but possibly somewhere in the lower crust or upper mantle, melting took place, forming magma. This magma could then have moved into the upper crust along zones of weakness such as folds, faults, and fractures, forming granitic bodies like the Bingham and Last Chance stocks. Hydrothermal solutions carrying enormous amounts of base and precious metals might have followed, forming the ore bodies.

This simple model outlines the events that could easily have produced the ore bodies in the Bingham copper-gold and associated mining districts of the Oquirrh Mountains. George Kennedy (1959), Gordon MacDonald (1963), and Roberts (1968) have discussed melting due to loading of the crust.

In 1967, with the help of Max Crittenden, Ed Tooker, Hal Morris, and Tom Cheney, I put together a paper on the Oquirrh Basin. This paper summarized our fieldwork, and the work of many others, giving us a broad picture of the eastern Great Basin. I was invited to go to Northwestern University, in the spring of 1968, to discuss the history of this region and integrate it with the geological history of central and eastern Nevada. The essence of my lectures was published in a paper on the "Evolution of the Cordilleran Geosyncline" in the *GSA Bulletin* in 1972. Unfortunately geosynclines were going out of style at that time, because they did not have clear-cut mechanisms to cause folding and thrusting. I pointed out, however, that sea floor spreading could be combined with geosynclinal deposition to form fold-and-thrust belts. I did not, however, emphasize "oceanic tectonics" as much as others preferred. Today, I would combine both mechanisms in a tectonic model of the western cordillera.

When I reluctantly left the Oquirrh copper-gold project, I chose a new challenge—to look at ore deposits in the Arabian shield. I had no idea what the future held but thought that I could carry on a project there with the same confidence that had served me well in Nevada and Utah.

Chapter 10

Life in Saudi Arabia

1971–1978

Oh! that the desert were my dwelling place.

—BYRON, *Childe Harold's Pilgrimage*

Saudi Arabia

In December 1971, I joined a thirty-man USGS mission in Jiddah, Saudi Arabia. There were also British, French, and Japanese teams, but ours was the largest group. The French were assigned to study ore deposits and map the northern part of the Arabian Shield, and we were assigned to study and map the southern part. Our team had preceded the others into Saudi Arabia; Roy Jackson went first and carried out a study of magnetic deviation throughout the country. Glenn Brown followed him and set up the mission, which, by 1971, involved thirty men.

It may sound strange for the USGS to have active projects in a foreign country, but actually, it was and still is a common practice. Saudi Arabia, lacking the geological expertise to study its own mineral resources, looked to established groups for help. This service not only benefitted the Saudi government, but it also helped the USGS, which received a bonus over and above its operating costs. This money could be used to fund important work back in the United States, and USGS geologists welcomed the opportunity to study the geology of other parts of the world.

Map 10.1. *Saudi Arabia and adjacent region. Note the incense route from Yemen to Gaza, a Mediterranean port now in Israel (black dotted line).*

I was in Arabia specifically to replace Conrad Martin, an economic geologist whose five-year contract was expiring. My duties were to study and describe the ore deposits. I had no idea that I would work on a treasure of monumental proportions—the Mahd adh Dhahab gold mine. This mine proved not only to be a workable mine, but was identifiable with the biblical mine Ophir, the source of King Solomon's gold.

Arriving in Arabia

Having worked in the American West, I thought that I knew the desert, but I didn't know the Arabian desert. I soon learned that there is no sharp seasonal weather pattern. The rain comes when conditions are right, not before. The people, camels, sheep, and goats hope for rain, but they cannot count on it.

Another attraction of Saudi Arabia for me was that I had roots there. I had learned from my nephew, Fred Norton, who researches genealogy, that our family had deep roots in the Middle East. He had traced our family line through the caliphs of Spain to the Prophet Muhammad through his daughter, Fatima. Our ancestors had been scattered from Mongolia to Morocco, Saudi Arabia to Iceland, and all points in between. So Arleda and I were fascinated by Saudi Arabia.

In the early 1970s the Oquirrh project had begun to slow down. We were not being allowed access by Kennecott to some of the deep-level data, so I turned the project over to Ed Tooker. He was perfectly capable of carrying on the project by himself, and he took it in a new direction, mapping the geology and structure of the northern Oquirrh Range, which did not interfere with Kennecott's interests. Ed has carried on this work in magnificent style. I had a few good years left before retiring, and I sought a new challenge. The boys were practically grown and would soon be on their own. I had always wanted to see the Middle East, and I was especially intrigued by reports I had heard about geologic work being carried on in Saudi Arabia. So I wrote James (Jim) Norton, chief of the Saudi Arabian project, and asked him whether he would accept me in Saudi Arabia. He was gratifyingly enthusiastic about my joining the project.

Jim planned to take leave and go home the summer of 1971, and he invited Arleda and me to join him and his wife, Kay, in South Dakota to discuss the possibilities. We met near Rapid City, discussed various projects, and the deal was closed. I was to join him in Saudi Arabia near the end of the year. I saw the opportunity as an incredible career-end bonus.

Arleda and I went to Jiddah in December 1971. Arleda and I were fascinated by Saudi culture and customs and charmed by the Saudi people, but we were overwhelmed with the feeling that we had traveled back to another century. We got off the airplane at Jiddah Airport and were immediately confronted by pure bedlam; hordes of people were everywhere, some waiting and others rushing. The noise was deafening. Women were covered in long black *obayas* and veils (*hijabs*), and the men wore *thobes* that nearly brushed the ground and covered their heads with *ghutras*. Some men fingered prayer beads and murmured Koranic verses softly to themselves while waiting. Arleda and I tried to wend our way through the hordes, looking for the proper line to enter for our passport inspections. We couldn't communicate; it was impossible to hear anything above the unintelligible shouting that seemed to come from everywhere.

We finally reached the line that seemed to be for foreigners; our assumption was confirmed when the immigration officer asked for our passports, visas, and proof of our religious affiliation. Both were carefully checked. The Saudis did not care which religion we practiced, but they did want to be sure that we had a religion—no atheists were allowed.

Next, we had a seemingly interminable wait while our luggage was meticulously searched, then searched again. We had been warned that the Saudi immigration officers would be looking for a variety of forbidden items, including drugs, alcohol, religious propaganda, and pornographic magazines, which were considered threats to the integrity of Islamic culture. Even bottles of prescription medicine were carefully scrutinized to make sure they did not contain contraband.

After all of our baggage had been checked and marked with chalk as proof of inspection, we were able to signal our friends, who were waiting for us outside the barrier. We were greatly relieved to be in their care, and they whisked us off to the USGS compound.

We were dropped off at our new home, which was in the "seven-house compound," so called to distinguish it from the "five-house," "the Zeban," and other compounds. Henry Rosario, a fellow employee, had thoughtfully gotten us enough food for a snack and for breakfast the following day. The USGS had arranged for a car and driver to be furnished to Arleda one day a week, from 9:00 A.M. until about 1:00 P.M. Arleda learned quickly that women were not permitted to drive, or to do a lot of other things, in Saudi Arabia. But that was part of the mystery and the charm of living in the Middle East.

On the arranged day, the driver would take Arleda to the *suq* (or *souk,* the market), where she could find a delightful choice of vegetables. Meat was generally quite lean and thus quite tough. Beef, lamb or mutton, goat, and even camel were available, although we never tried the latter. Normally, the butcher cut each muscle separately from the hindquarters, so one had to know the cuts of meat well. Arleda opted for the tenderloin whenever she could, even though the market's chickens were good, and the fish was excellent. The fish market was filled with many strange varieties, among them something called nadju, which we grew to love. Barracuda, which weighed from fifteen to twenty-five pounds, were prized for special buffet dinners.

Safety was a concern, but we quickly learned to sterilize our vegetables and fruits in boiled water to which an ounce or two of bleach had been added, and as an extra precaution, we boiled all the water we drank or used for cooking.

Our house was a one-story brick bungalow with four bedrooms and was equipped with modern appliances, including an electric refrigerator, a gas stove, and, to our good fortune, a large air conditioner, which was built into the house and which maintained it at a constant, ideal temperature of seventy-six degrees Fahrenheit.

As we had seen at the airport, the *thobe,* a garment that nearly reaches the ground, is the usual dress for men. In the winter it is made of closely woven and relatively heavy material, but in summer it is light, almost diaphanous. Under the *thobe* a man wears undershorts that nearly reach the knee—the length prescribed by religious law.

During the 1940s, Prince Faisal al Saud, as a young delegate to the UN Conference in San Francisco, took to wearing briefs as Americans do. It is reported that once he was wearing briefs with large polka dots, when he realized that they showed prominently through his thin, summer-weight *thobe.* He changed back to his customary undershorts quickly.

The *ghutra,* or headcloth, is commonly white in summer and checkered in winter. The *ghutra* is held on by an *egal,* originally made for hobbling camels but now adapted for stabilizing the headdress. *Ghutras* also make excellent tablecloths, and one is a perfect fit for a bridge table. Women always wear long dresses with long sleeves and, of course, veils when in public to keep men other than their husbands or fathers from gazing upon them.

After we had been there a few weeks, we moved to the five-house compound, where Jim and Kay Norton lived. Because USGS assignments in Saudi Arabia were generally for two-year periods, housing arrangements were like a game of musical chairs, with some people leaving just as others arrived.

The distinct differences between Middle Eastern and Western cultures could have made social life difficult, but the USGS mission adapted well. We continually socialized with members of our own and other compounds, even including people from the British and French geologic missions.

In the Field

The landscapes of Saudi Arabia are varied, ranging from the relatively flat coastal plains that border the Red Sea and Arabian Gulf to the Hejaz uplands that rise from four thousand to nine thousand feet in altitude. These uplands are dotted with *jabals* (steep hills or peaks) separated by broad alluvial plains that contain the highways of the upland; in places twenty or thirty overlapping auto tracks can be seen when one flies over the Hejaz. The Hejaz has long been crossed by major trade routes from Yemen to the port of Gaza on the Mediterranean. There are many ancient trails in Saudi Arabia, such as one along which incense was traditionally carried. Sand dunes cover parts of the Hejaz, but the major dunes are in Ar Rub' al Khali (the Empty Quarter) in southeastern Saudi Arabia. The higher parts of the Hejaz, those above eight thousand feet, receive eight to ten inches of annual rainfall and contain perennial streams. In places some snow falls in the winter. In the southwest, near Jizan, temperatures can reach above 120 degrees Fahrenheit, but fortunately, it was never that hot in Jiddah.

Our geologists found it impossible to work in Jizan in the summer. This was a good excuse to explore the northeast margin of the Red Sea and its fabulous coral reefs with their clams and colorful fish, many at snorkel depth. We would rent the embassy boat from time to time, going out in the morning to a point about six miles offshore, snorkeling until we had our fill, then resuming for a few hours after lunch.

Whenever we set up camp in the desert, our Muslim coworkers would first erect a low wall of rocks pointing toward Mecca to guide them when they said their prayers five times daily. One Muslim would be designated to lead the prayers, and he would kneel, facing Mecca, and proceed with the prayers. The others would then do just as he had. I continue to be impressed with the Muslims' total commitment to their religion; one does not often see signs of backsliding.

Our relations with our Saudi crews were cordial, and in the evenings we often shared after-dinner *gawha* (coffee) and *chai* (tea), alternating them for an hour or so.

Once, during my travels in a remote area of the kingdom, I witnessed a scene that will always be a vivid memory. A group of us geologists were in the southeastern part of the Arabian Shield; near day's end, we began to look for a goat for dinner. We had heard of a farmer who had goats to sell, so we drove to his farm to buy one. Five of us were in a Land Rover, and we parked near his house. Our Arabian guide went to find the herder and buy the goat. While we were waiting, two veiled women emerged from the house. Both were wearing long dresses, one a deep tangerine color and the other a brilliant shade of green. They stopped about forty feet away from us and began conversing. They acted as though they were unaware of our presence, but I suspected that they had donned their best dresses just to impress

FIG. 10.1. *Enjoying tea (*chai*) at field camp. Rear (left to right): Kenneth McLean, Frank Dodge, Dwight Schmidt, M. Naqvi Mustafa Mawad, and Gerald de Roux, Saudi Arabia, ca. 1972.*

us. They remained there, seemingly posing, for at least fifteen minutes. I badly wished to photograph them, but taking pictures of Saudi Arabians, especially women, was risky. Most Saudis would react quite strongly when a camera was brought into their sight. These women looked as though they wanted us to photograph them, and perhaps they did; I'll never know.

They moved away as our guide returned with a very angry goat, which he put into the back of the Land Rover. When we reached camp at dusk, the goat was slaughtered in the Muslim manner: its head was pointed toward Mecca, and its throat was slit. It was held down until it finished bleeding.

After a ritual slaughter, goats (or sometimes sheep) were skinned, eviscerated, and put into a large pot to cook with rice. This was a time-consuming method of cooking, and dinner was served quite late in the evening. I had thought that goat meat or mutton, especially from a mature animal, would be tough and quite odorous. But, when cooked for three or four hours with rice, ghee, and proper seasoning, it was tender and tasty. We enjoyed these dinners, though most of us refused the eyeballs, which were always offered to us.

Working in the field was always fascinating to me because the geology and the people of Saudi Arabia were so new to me. I never got used to the fact that wherever we landed with a helicopter, within five minutes, an Arab sheepherder would appear and invite us to tea at his camp. If we accepted, an invitation would be issued for lunch or dinner. Had we accepted every invitation we received, we would never have been able to complete our work, so generally, we refused invitations with a cordial *shukran* (thanks).

Arleda sometimes could accompany me into the field. She felt perfectly at home conversing in her limited but effective Arabic.

I was constantly watching for an opportunity to take some photographs. Once we hired a Bedouin to guide us to an old mining site. When we took him back to his house after the trip, I finally got my opportunity to photograph a family. The Bedouin freely offered to let us take pictures of his wife and daughters. They were washing a sheep to take to market, and we photographed them without veils. They must have enjoyed it, too, since they giggled the whole time we were shooting.

One summer, we were working in the southern part of the Arabian Shield. That year (1976) water was especially scarce in the Wadi Wassat area, and a Bedouin family asked if we could spare some drinking water. When we said we could, they filled several large truck inner tubes with water and loaded them onto a camel and a burro. Then, like the Bedouin guide, the husband offered his gratitude and said that we could photograph his wife and child. I obtained several striking photographs of the family. Their son, who looked to be about eight years old at the time, gave me a shy smile as I took his picture. The man's wife, however, was very solemn behind her veil.

Another time, my guide, Ghanen Gheri, and I drove into the backcountry and came upon a family moving their stock from low-altitude winter pastures to high-altitude summer pastures. As we came up to them, we noted that they were watering their camels, sheep, and goats from a deep well. We approached the well, and the men greeted us with the familiar "Salaam aleikum," and we responded with the customary "Wa aleikum salaam." They were very cordial to us and flashed broad smiles. We could see that they would be engaged for a long while in watering their stock, for the water had to be drawn up about sixty feet. One man climbed down onto projecting stones left in the wall and dipped up about three gallons at a time, and then the men on the surface pulled it up and poured it into a trough for the animals. They told us that we could photograph the men, their sons, and the stock, but not the women and girls who were standing nearby. We, of course, agreed, and I obtained some excellent shots of camels, sheep, and goats waiting patiently to be watered. No pushing and shoving allowed!

On another occasion, we were driving along the foot of the Hejaz, when we saw another family going to the upland summer grazing lands. The women were traveling in shaded camel-borne *maksar* (palanquins), which swayed precariously as the camels moved over the rough ground. The women were veiled, but it was obvious that they were young. I was not able to take any photographs that time, but the image has lingered in my mind because it seemed like a scene out of the distant past.

Not all my photo opportunities involved the common people. One winter day, I was working in Wadi Bishah, in southern Arabia, with Conrad Martin, and we received an invitation to visit the camp of Prince Faisal al Saud, the son of King Faisal. Conrad told me that this was not a casual invitation, but was a command, as the prince was also governor of Abha Province, the site of the USGS project we were visiting. The next morning, we asked our helicopter pilot to drop us at the camp about 8:30. The prince was being shaved, and so we waited for him for half an hour. Once he was ready, he greeted us cordially and asked us to sit down and take coffee with him. The coffee was prepared in the traditional manner, by first roasting the beans lightly, then grinding them along with cardamon seeds to a fine powder, then steeping the powder in boiling water. The coffee was then served in tiny cups. When we had each finished two cups of the coffee, the prince asked that tea be brought. The tea, sweetened, was served in small glasses. The prince next asked for coffee again, and then tea, then more coffee. We finally indicated that we had had enough by turning our cups upside down. Then the prince asked what kind of work we were doing in the area, and we told him that we were evaluating the economic potential of surrounding mineral deposits. We said that we had

Fig. 10.2. *Visit to field camp of Prince Faisal al Saud (left to right): retainer, author, two retainers, Prince Faisal al Saud, two retainers, Conrad Martin, and two retainers, 1972.*

not found anything commercially significant as yet in the Abha Province but that we would let him know if we did. Then he asked if we wanted to photograph him and his retinue, and we eagerly agreed, so he and his men lined up and allowed us to take as many photographs as we wanted. Most of his retainers wore gold-plated daggers and looked very impressive.

Sometime later, when Conrad and I were in Abha, the provincial capital, we visited the prince again and updated him on our activities. Conrad mentioned that we had recently visited a green meadow in which a perennial stream flowed—a rare sight in Saudi Arabia—and allowed as how this area would make a fine park. The prince was horrified! He said that the area belonged to a certain tribe, had always belonged to that tribe, and would always belong to the tribe. Moreover, one did not take water away from its owners—they might die of thirst. The concept of "eminent domain" was evidently not part of Koranic Law. (I understand that now some land in the high country is being set aside for parks. That represents quite a change from the thoughts then expressed by Prince Faisal al Saud.)

Once in Bahrain, I noticed an Arab carrying a hooded saker falcon on his arm. I asked him (in English) if he would permit me to photo-

FIG. 10.3. *Saker falcon and handler, Dubai market, ca. 1975.*

graph the bird. He nodded then unhooded the falcon so that I could take a really good photograph. As always when I was allowed to take a photo, I was very grateful, in this case, to both the man and the falcon!

The Saudis are great falconers. It is quite common to see trucks with perches in the rear compartments, each carrying four or five hooded falcons perched on a bar, returning from a hunt. Their game in some areas is the great bustard (a large bird related to the crane), which is nearing extinction. More often their game is the lesser bustard or the See-See partridge.

The Saudi Arabian culture became a part of our camps. The equipment used varied with the location, whether one was on the sizzling coastal plain near Jizan or in the relatively cool Hejaz. On the plain, Portacamps with cooling units were available. A generator provided electricity for air-conditioning, lighting, and the radio. In the higher elevations of the Hejaz, tents normally sufficed during the winter months. We customarily took a cook, usually Yemeni, a couple of camp hands, and a guide. The guide was generally from a local tribe, knew the people, and could deal with any problem that came up. He would accept, with our permission, invitations from the local amir and would extend invitations to the amir when appropriate. Mostly, local tribes were hospitable, for they were advised that our work might result in employment for their people, but not all tribes welcomed strangers. In those cases, we weighed the consequences and avoided tribes that seemed inhospitable.

Our cooks provided plain, wholesome meals. Some even learned to bake pies and cookies. The project chief usually bought the food, splitting the cost with his assistant, if he had one. This way, the project chief could choose food he liked, though if the assistant and the pilot had preferences, they were accommodated. (The helicopter pilot assigned to the project, if there were one, also shared the cost of supplies.) The project chiefs, assistants, and pilots received adequate money for field expenses. The Saudi camp workers and guides furnished their own food and prepared their own meals (they were also given per diems).

We generally traveled to the field by airplane; our equipment was sent by truck and accompanied by workers who set up camp. At times I traveled from base camps—with airstrips and fuel supplies—to remote sites by helicopter. We essentially had our own little air force, piloted by British, Canadian, and American pilots for the most part, all of whom were versed in the art of bush flying. I particularly remember three pilots: Gerry de Roux, Ray Jackson, and Boyd Shaw. Even though the airstrips were small and rough, I never felt any hesitation in flying with these men. They would never fly in unsafe conditions, such as dust storms or thunderstorms.

Storms were not the only hazards in Saudi Arabia. The country has several varieties of poisonous snakes as well as many scorpions. The snakes include cobras with narrow hoods; sea snakes in the Red Sea, which are reported to be very venomous; and a small viper, which is only six to eight inches long but has deadly venom. I saw one of these vipers in the field, after it had been killed by some Saudi workers. I commented, "That is only a baby!" They assured me, "one bite and you are in Paradise." It seems likely that Cleopatra was killed by such a viper, although on TV it is commonly designated as a cobra. In 1977, a USGS hydrologist was walking in the desert at night when he felt a sharp prick, like that of a thorn. He ignored it, but by the next morning he was in great pain. Unfortunately, it was a weekend, and no one in Jiddah answered the radio call for a rescue plane. He died later that day. I decided not to ever again wander out into the desert at night. The only cobra I ever saw was a dead one; they are nocturnal. Not all of Saudi Arabia's creatures were dangerous; we sometimes saw hedgehogs out in the field. They are cute little mammals related to shrews and moles. They are now extinct in North America, but about fifteen million years ago they lived in Nevada.

The highlight of my Arabian tour was seeing old gold mines, most of which had been productive on a small scale, and one in particular—Mahd adh Dhahab—that will shape the future of mining in Saudi Arabia.

Chapter 11

Ophir

MAHD ADH DHAHAB, "CRADLE OF GOLD"

> Discovery consists of seeing what everybody has seen and thinking what nobody has thought.
>
> —ALBERT SZENT-GYÖRGYI VON NAGYRAPOLT, *The Scientist Speculates*

Soon after I arrived in Jiddah, Conrad began planning trips to show me some of the ore deposits. We first went north to visit the projects being carried on by the French mission, who were studying Precambrian rocks. We were flown to the northern part of the Arabian Shield in a De Haviland Beaver piloted by one of the best of the bush pilots employed by the Saudi Arabian Ministry of Petroleum and Mineral Resources, Gerald de Roux.

In early March 1972 Gerry took us first to Nuqrah, the BRGM (Bureau Recherche Géologiques et Minières) lead-zinc-silver project. Next we went to Jabal Sayid, a copper project. I enjoyed visiting the site, not only because of the unique geology and ore deposits, but also because the French had good cooks who prepared excellent meals. On this trip we stayed the night because dinner was finished after dark, and our planes were not allowed to fly unless their pilots could clearly see the ground.

Reconnoitering Jabal Mahd adh Dhahab

The next morning, as we flew back to Jiddah, we passed near an old gold mine at Jabal Mahd adh Dhahab. As we approached Mahd adh Dhahab (Cradle of Gold) I saw a mind-boggling sight. A series of fifteen or twenty parallel quartz veins were clearly visible crossing the jabal from north to south. I asked Gerry to circle the jabal once, then twice. After completing the second circle, he asked if I wanted to land. I replied, "YES" emphatically. Not only had I seen the ancient open stopes (openings along a vein from which ore has been removed) mingled with modern workings (SAMS—Saudi Arabian Mining Syndicate—1939–1954) in the principal mineralized area on the northeast, but I had recognized two other areas of ancient workings, one to the west and another to the southeast of the jabal. I could also clearly see that neither site had been explored in modern times. This told me that the ancients had found some gold in the old pits, but that the ore was lower in grade than that they were mining elsewhere. I surmised that high-grade ore might underlie the low-grade surface ore.

Gerry managed a very bumpy landing on one of the alluvial fans on the east flank of the jabal. When we all jumped out of the plane, Gerry examined the tires and saw a four-inch gash in one of them that exposed the inner tube. Gerry said, "Whew, we were lucky it didn't blow!"

Conrad and I scrambled up the jabal, visiting first the ancient open stopes and modern workings in the northeastern area, then continuing to the western and southeastern sides, where ancient workings had not been disturbed in recent times. The extent of metallization in and around these pits was most impressive. In addition, Conrad and I noted abundant fragments of copper minerals such as malachite (green copper carbonate) around the pits and along quartz veins.

I told Conrad I thought the mine might have a long way to go before its potential was exhausted. I resolved to revisit Jabal Mahd adh Dhahab and take samples of the ancient workings as soon as possible. Then we returned to Jiddah. We had no spare tire, and there was no way Gerry could repair the damaged one, so we hoped for the best as we took off from the rocky flat. It held together.

FIG. 11.1. *Aerial view of Jabal Mahd adh Dhahab, workings from 1937–1954 and ancient stopes on left.*

Before Conrad left Saudi Arabia to return home, I accompanied him on several more trips to visit ore deposits in the Arabian Shield, but none had the potential of Mahd adh Dhahab.

I returned to Mahd adh Dhahab in July with Robert Luce, Frank Dodge, Dwight Schmidt, and a Saudi guide, Ghanem Gheri, and I asked each of them to sample the veins in a small area along the southeastern side of the jabal, where Conrad and I had seen ancient workings. I sampled another segment. We collected in all about fifty grab samples (samples collected from loose rock fragments on the surface); twelve of them yielded 0.35 troy ounces or more gold per metric ton (tonne). So, armed with these excellent values, I proposed to Thor Kiilsgaard, our new chief of mission, that we map and evaluate the area for an exploration project. Thor agreed and assigned Robert Luce and me to the project, but since the project was in the area controlled by the French mission, we first asked the French whether they wanted to work at Mahd adh Dhahab. They had no plans to do so, because they were deeply involved with a copper project at Jabal Sayid, nor did they object to our working the site. Bob wanted to gain experience in geologic mapping; therefore, he carried out the initial stage, and I checked his work periodically. Mapping was started in late 1972, to avoid the heat of summer and fall. We were joined by Abdulaziz

FIG. 11.2. *Abdulaziz Bagdady at ancient open stope, Mahd adh Dhahab.*

("Ziz") Bagdady, a geologist from the Ministry of Petroleum and Mineral Resources, to help carry on the multiple projects we planned. Bob, Ziz, and I conducted an early geochemical sampling program in the area to learn more about the distribution of gold, silver, lead, zinc, and copper. We identified several positive anomalies of these metals and marked them for further study (Roberts, Bagdady, and Luce, 1978).

Our camp at Mahd adh Dhahab was set up on the southeast side of the jabal. A short landing field was leveled near the camp. In addition to our cook, we had two men who did camp chores, purchased supplies, and took care of local business.

Some social activity was interspersed with our mapping and sampling program. Arleda was able to travel with us into the field, and we were once invited to dine with the local amir, Abdullah Rahman Assudieri, the political head of the province. Bob, Ziz, Arleda, and I arrived, and Arleda was promptly shown to the women's quarters and we to the men's quarters. The amir hosted a traditional dinner; a whole sheep, liberally saturated with *ghee* (butter), was served on a bed of rice. I was offered the sheep's eyes, which I declined as graciously as possible, but I enjoyed the rest of the meal. We not only had to adjust to traditional food but also to the traditional way of eating. One reclines against pillows, reaches out and grabs a handful of

FIG. 11.3. *Al Amir Abdullah Rahman Assudieri (left) and Bob Luce at field camp, Mahd adh Dhahab.*

rice, always with the right hand (because the left hand is considerd unclean), kneads it into a ball, then pops it into one's mouth. Commonly the amir, or one of his men, will hand the guests slivers of meat that are carved as needed. This goes on until everyone has eaten his fill. Afterward, pitchers of water and towels are offered so that diners may wash and dry their hands. Fruit, generally an orange or an apple, is served, then water is again brought in for a second rinsing of hands, and a towel is again furnished for drying the hands. Arleda rejoined us after dinner, and we returned to our camp. Arleda's Arabic was good enough for casual bartering or getting directions but not for more formal dinner conversation. She reported that a teacher who spoke a little English was present at the women's dinner, so she was able to communicate adequately with the ladies.

We reciprocated with a dinner invitation shortly thereafter. We set up a large tent, about thirty by sixteen feet, and laid down coarsely woven Iranian rugs. The effect was elegant, and I had the sense we might have stepped onto the set of a Hollywood movie. We also served the traditional sheep on a bed of rice and extras, but we did not separate the men from the women; Arleda served graciously and successfully as hostess. One of our staff assistants, who was a member of

FIG. 11.4. *Ancient anvil and hammer for crushing gold ore found at Mahd adh Dhahab.*

the local tribe, acted as interpreter. The amir spoke minimal English, mostly as a courtesy to us. He was happy that we had come into Mahd adh Dhahab to review possible reopening of the mine. That part of Saudi Arabia had been hard hit by drought, and jobs were needed for the townspeople. Some did get jobs working with foreign geologists or engineers, and they all were hoping that mines would be opened to give more permanent work.

Day after day, as we worked, we would notice herds of goats and fat-tailed sheep grazing along the slopes, gleaning the new shoots of acacia trees, which were struggling to keep alive. A tiny shoot did not last long with those voracious animals. I felt sorry for the plants—never getting a chance to grow.

Examination of old workings showed us that ancient miners had operated quite efficiently. In the old part of the mine we found hammers, anvils, and other crushing/grinding tools that gave us a good idea of the methods they used. They first removed the rich surface ore, then mined the upper parts of the veins in stopes. After that, the ore was crushed on an anvil, coarsely ground in a circular quern, and finely ground in a stone with a trough cut into it. Then the resulting powder was panned to separate the gold and silver from other minerals, such as quartz and iron oxides.

FIG. 11.5. *Circular quern for coarse grinding and trough for fine grinding of Mahd adh Dhahab ore.*

As our investigation proceeded, it became evident to Bob and me that exploration was warranted. Samples from the ancient workings showed significant values in gold and silver and traces of copper. I suggested that the most accessible part of the vein system, in the southeastern part of the project area, was the obvious place to drill; we could then explore extensions of the veins that had been mined in shallow pits dug by the ancient miners on the jabal above.

Bob, Ziz, and I invited Thor to the site and made our case. Thor gave us permission to go ahead. To formally set up an exploration project, the mission concerned went before the exploration committee at the Ministry of Petroleum and Mineral Resources and presented justification for the project. Bob and I presented our proposed plan of exploration, emphasizing our excellent assay data on surface samples. Our project was quickly approved by the committee.

Drilling Project at Mahd adh Dhahab

The first two drill holes were initially laid out by Bob, Ziz, and me. These cut low gold and silver values, indicating that we were on the right track. Then Ron Worl, from Denver, Colorado, came to Jiddah, and we turned over the project to him. Ron modified the program

FIG. 11.6. *Ronald G. Worl, Mahd adh Dhahab.*

(Worl, 1978), choosing a spot for the third hole 150 meters northeast of the first two in order to intercept a favorable unit (lower agglomerate—coarse-grained volcanic rock), near the contact of the overlying lower tuff (fine-grained volcanic rock). John Kemp and others (1982) have assigned these units to the Has Formation of the Mahd group. In July 1999, Ron Worl informed me that Gavin Dirom, a former employee of SAMS, had reported in 1954 that high gold values occurred just below the upper tuff. The third hole cut impressive gold values at a depth of 174 to 182 meters, containing more than an ounce of gold per tonne (Worl, 1978, p. 30). These values constitute a valid discovery that Bob Luce, Abdulaziz Bagdady, Ron Worl, and I shared. Worl drilled three more holes that did not contain significant gold, but the seventh contained excellent gold values, permitting him to calculate that Mahd adh Dhahab "has a potential resource of 1.1 million metric ton[ne]s containing 27 grams [nearly 1 ounce] per ton[ne] gold and 73 grams [2.3 ounces] per ton[ne] silver" (Worl, 1978, p. 2).

This had been precisely the right place to drill; with only seven holes, Ron Worl came up with reserves warranting construction of a mill. At this time gold was worth $700 an ounce, so we were contemplating $700 million worth of gold. This resource certainly indicated a workable ore body, even in a high-cost country like Saudi Ara-

FIG. 11.7. *Lowell S. Hilpert on ruins of ancient house, Mahd adh Dhahab.*

bia, where operating costs are so high because of the extreme environment. Most surface exposures here showed secondary copper minerals, such as malachite and chrysocolla. Luce, Bagdady, Worl, and I sought and found golden treasure. It was just one more instance of "being at the right place at the right time." We were happy to have completed a viable project, one that justified the U.S. Geological Survey's many years of effort in Saudi Arabia.

Arleda and I returned to the United States in 1978. Then in 1982, Lowell Hilpert and I went to Saudi Arabia to map the northern part of Jabal Mahd adh Dhahab. Arleda accompanied us, but she remained in Jiddah while we were in the field. We completed the mapping and wrote a report on the old workings and surface geology. Lowell and I also visited the old townsite nearby. We hoped that one day this area could be excavated and studied by archaeologists, so that light could be shed on the early production of gold from Mahd adh Dhahab. It seems likely that datable material might be found there.

Gold Mining in 1989 at Mahd adh Dhahab

A new mine began operation in 1989 with a mill capable of treating two hundred and fifty to three hundred tonnes of ore daily. After a difficult start-up period, the mine and plant have operated smoothly. In January 1994, Ziz and I revisited Mahd adh Dhahab to see the modern operation of the underground workings and mill; the ore was then yielding about 0.9 ounces of gold per tonne, and about three hundred tonnes of ore were being treated daily in the mill. The reserves were about a million tonnes. When I was asked during the visit how I happened to find the new ore body, I simply replied, "we just explored the ground below the old pits where the ancient miners had worked. We knew that they would dig only where there was the promise of high-grade gold ore."

The mining operations of Mahd adh Dhahab have led to the discovery of new ore bodies of minable grade, and it is likely that additional discoveries will be made in the future. There may also be ore of a lower grade surrounding the rich ore, and this ore will likely be evaluated for possible production when the price of gold is favorable.

Ron Worl wrote me in February 1999 to report that Saudi Arabia had recently formed a national mining company, Ma'aden, under the control of the minister of petroleum and mineral resources, with an initial budget of two billion dollars. Ma'aden has taken over operation of Mahd adh Dhahab along with control of all gold mining and exploration in the country. Currently, Mahd adh Dhahab gives 1.1 million ounces of gold as its reserve figure. It looks like Mahd adh Dhahab will be a 5 to 6 million ounce deposit (a world-class ore body according to RJR) by the time mining is completed. Ma'aden has begun lowering the mine's mill grade to 0.55 ounces per tonne, which will greatly increase total production of gold. Mine geologists have shown that there are significant gold values across broad areas of altered rock between the major veins. In his letter, Ron went on to say, "Another gold project being prepared for production by Ma'aden is Al Hajar in Wadi Shwas . . . the nearby Jadmah prospect also contains a small body of gold-bearing oxidized ore."

In the course of our investigations at Mahd adh Dhahab, we attempted to obtain information on the early history of the mine. We were aware that Karl Twitchell (1958a), a consulting geologist, who had been furnished by the Crane Plumbing Company to the kingdom to aid in the search for water sources, had also looked at gold mines, among them Mahd adh Dhahab. He had been strongly impressed with the deposit and had suggested that Mahd adh Dhahab and the biblical Ophir might be one and the same. Twitchell, however, gave no data to back this up, and there was no hard proof that Ophir had ever existed. No mine capable of producing the enormous amount of gold attributed to King Solomon in 1 Kings had yet been found. I thought Mahd adh Dhahab was a candidate, and after five years of research, I found that only one mine in Saudi Arabia could have yielded that amount of gold—Mahd adh Dhahab. I have summarized all available data that led to this conclusion in the following account.

Mahd adh Dhahab, the Ophir of Antiquity?

One of the most fascinating tales of lost mines and great treasure is that of Ophir—a fabulously rich gold mine of biblical times said to have been worked by King Solomon and other area rulers. Every prospector dreams of discovering an Ophir. So throughout the world, and especially in Latin America and the western United States, the name "Ophir" has been applied to many prospects, but they rarely lived up to expectations. But the hope lingers of finding instant, practically unimaginable wealth.

Where was Ophir, the source of King Solomon's gold? Although Ophir is frequently mentioned in the Bible, its location is never described precisely. The Bible documents large-scale gold production during Solomon's reign, so it must have been a mine (or mines) of significant size with extensive workings.

How could such an important mine have escaped the attention of historians from 950 B.C. to A.D. 1932 and have virtually disappeared from literature? The historical record offers meager clues. Though several writers have guessed that Ophir might lay in Arabia (Burton, 1878; DeKalb, 1914; Rickard, 1932), no identification was made until Twitchell (1958a) tentatively linked the Mahd adh Dhahab Mine in Saudi Arabia to King Solomon's Ophir.

Some historians and archaeologists have questioned the very existence of King Solomon (Herzog, 1999; Niebuhr, 2000), while others (Dever, 1999) have affirmed Solomon's reign. Kenneth Kitchen (1989) has tied King Solomon to Egyptian history. It is not my role to judge this issue; in the following discussion I plan merely to evaluate evidence as to whether any mine in Saudi Arabia could have furnished the gold attributed to biblical Ophir.

The data I have assembled, my analysis of the historical record, and my evaluation of ancient mining operations at Mahd adh Dhahab all strengthen Twitchell's claim. I also refer to biblical accounts of Solomon's mining activities, though I recognize that they have not been validated by archaeological investigations. Again, the question is not whether the account in 1 Kings is factual, but, rather, whether Mahd adh Dhahab could have supplied the gold mentioned in that account.

In order to positively identify Ophir, the following requirements must be met: (1) the mine (or mines) must have been capable of yielding a large production of gold (34 tonnes according to biblical chronicles) during ancient times; (2) it must have been reasonably accessible by sea and by land from Ezion-Geber at the head of the Gulf of Aqaba; and (3) the mine must have been productive during King Solomon's reign (961–927 B.C.). After a look at the historical accounts relevant to Ophir, I will briefly outline pertinent information that meets these requirements.

The Historical Record

T. A. Rickard (1932) has prepared the most comprehensive summary of the historical fragments available in the literature relating to Ophir. He states that no confident reply could be given to the question "where was Ophir? . . . despite much search and even more controversy. Among the regions put forward for being the great gold field of antiquity are Arabia, Armenia, Ceylon, Haiti, India, Malaya, Peru, Phrygia, Rhodesia, Spain, and Sumatra."

> Modern investigations have reduced the principal claimants to two, Arabia and Rhodesia. In 1868 Renders discovered the ruins of Zimbabwe in southern Rhodesia. In 1871, Karl Mauch described these ruins. Twenty years later Theodore Brent and R. M. W. Swan examined them scientifically. Brent declined to identify Zimbabwe with Ophir but claimed that from this country the ancient Arabians got a great deal of

> gold and were in contact with both Egypt and Phoenicia before the Sabai-Hamaritic period. This statement was challenged by MacIver, who asserted (on the basis of Chinese pottery fragments discovered in the ruins) that the ruins were not older than A.D. 1400–1500. (pp. 242–43)

More recently Page and Oliver (1972) have dated older buildings at Zimbabwe at about A.D. 1000. If these are the oldest dates at the site, Zimbabwe must be eliminated from the competition as a potential site of Ophir.

Rickard continues, "Let us now consider the claims of Arabia" (p. 255). As early as 2450 B.C. in the days of Gudea, governor of the city of Lagash, gold mined in Arabia was sent to Mesopotamia via Dilmun (thought to be the current Manama, Bahrain), a port on the Arabian Gulf; this gold came from the drainage basins of the Wadis Rumma (now Ar Rumah) and ad Dawasir in central Arabia and was shipped via the oasis of Yabrin (also Jabrin) (Lutz, 1924). Dilmun, on the north side of the island of Bahrain, was the major trading center through which commodities, especially copper, ivory, and spices, were shipped to Mesopotamia during the second and third millennia B.C. (Bibby, 1969). It is also possible that gold sent to Lagash passed through Dilmun.

Although Rickard searched all available Western sources, he found only vague references to Ophir. Many accounts were outrageous speculations that only further confused the issue (Burton, 1878; Keane, 1901; Crawfurd, 1929). Richard Burton (1878) shrewdly guessed that the site of Ophir was in Arabia, but in 1910, after a fruitless search, he broadened his search to include other parts of the Middle East, India, and North Africa. With so little reliable information to go on, Rickard finally expressed doubt about the existence of any gold mine in Arabia capable of meeting the requirements of biblical Ophir. In retrospect, it is not surprising that Burton and Rickard were unable to locate Ophir; no one had made a thorough survey of gold mines until the American mining engineer K. S. Twitchell carried out his explorations in the 1930s. Rickard's account was published in 1932. No doubt he was putting the finishing touches on his manuscript just as Twitchell was beginning investigations that would again raise the question—where was Ophir? A map of Saudi Arabia gives no clue as to its location, but it does show the incense route from Yemen to Gaza (see map 10.1). The mine lies within view of the incense route, two miles from Bir Ma' aden, an oasis on the trail.

The Modern Historical Record

During his investigations, Twitchell noted ancient mine ruins and tailings in many places. He reported these sites to the king, who asked him to evaluate them for possible exploitation. Most of the ancient mines proved to have only limited potential, but Mahd adh Dhahab showed great promise, and even seemed to Twitchell to likely be the famous Ophir.

Twitchell summarized his first impressions of Mahd adh Dhahab as follows: "From the general appearance of the ancient stopes, as well as the tailings, it seems not impossible that these date back to King Solomon . . . the workings of Mahd adh Dhahab are the largest I saw in Arabia . . . [and] it is a reasonable guess that this [mine] might have been the source of the gold of King Solomon" (1958a, pp. 247–48).

Though Twitchell did not specifically mention Ophir, there is no doubt that he was referring to it in his 1932 account. He therefore deserves the credit for first linking Mahd adh Dhahab and Ophir. His "reasonable guess" needs further elucidation and clarification, which can best be done after a discussion of the biblical claims for the size of Solomon's treasure. In 1991 Time-Life Books reported Twitchell's conclusions as to the identity of Mahd adh Dhahab and Ophir in the volume *Lost Treasures* (1991, p. 18), noting that USGS work by Hilpert, Roberts, and Dirom (1982) reinforced Twitchell's conclusions.

The Biblical Record

The principal characters in the following account are King Solomon, King Hiram of Tyre, and the Queen of Sheba. King Solomon, son of King David, ruled Jerusalem from 961 to 927 B.C. King Solomon was associated with a contemporary Phoenician ruler, King Hiram, in many business ventures, among them voyages to Ophir to obtain gold and other valuable materials. The Queen of Sheba from Marib,

Yemen, was a famous visitor to King Solomon's court. Much has been written of this visit (DeKalb, 1914; Keller, 1956; Pritchard, 1974). Although the visit is not directly related to the story of Ophir, it does throw light on the life and times of those days and helps place Ophir in a realistic perspective. M. B. Davidson (1962) also portrays events of this period.

The Bible mentions Ophir in Genesis, Exodus, 1 Kings, and Chronicles. The most pertinent passages from 1 Kings appear below:

1 Kings 4:21

Solomon [. . . King over all Israel . . .] ruled over all the Kingdoms from the Euphrates to the land of the Philistines and the border of Egypt.

1 Kings 6:1, 20, 21

So Solomon built the house [of the Lord] and overlaid the inside of the house with pure gold and he drew chains of gold across, in front of the inner sanctuary and overlaid it with gold.

1 Kings 7:48–50

So Solomon made all the vessels . . . the golden altar, the golden table . . . the lampstands of pure gold . . . the flowers, lamps and the tongs of gold.

1 Kings 9:14

Hiram had sent to the King one hundred and twenty talents of gold.

1 Kings 9:26–28

King Solomon built a fleet of ships at Ezion-Geber, which is near Eloth on the shore of the Red Sea . . . and [King] Hiram [of Tyre] sent seamen together with the servants of Solomon and they went to Ophir and brought from there gold to the amount of four hundred and twenty talents . . . to King Solomon. [Weights in ancient times in the Middle East were initially based on Babylonian standards. The talent used throughout the region ranged from 25.2 to 30.43 kilograms (about 63 pounds); Scott (1959) favored the talent of 28.53 kilograms, and that unit will be used here. It is divided in 50 minas of 570.6 grams each or 2,500 shekels of 11.41 grams each; a talent contained 917.27 troy ounces.]

1 Kings 10:1, 3, 10

Now when the Queen of Sheba heard of the fame of Solomon she came to test him with hard questions . . . and Solomon answered all her questions . . . and she gave the King a hundred and twenty talents of gold [120 talents—110,072 troy ounces—3,424 kilograms].

1 Kings 10:14–17

Now the weight of gold that came to Solomon in one year was six hundred and sixty-six talents . . . and Solomon made two hundred large shields of beaten gold; six hundred shekels went into each shield and he made three hundred shields of beaten gold; three minas of gold went into each shield.

These figures indicate that Solomon had substantial amounts of gold at his disposal. Not all of it had to have come from Ophir, but a large part was probably derived from Ophir or nearby mines.

Sources of Gold

The sources of the Queen of Sheba's gold may have been small gold mines known in southern Arabia, Yemen, and nearby countries. Over a period of time these could have collectively contributed the stated 120 talents (about 3,424 kilograms). Also, the queen may have obtained gold in exchange for myrrh and frankincense from her kingdom in Yemen. Gus Van Beek told me in 1974 that the incense route from Yemen to Gaza on the Mediterranean (map 10.1) was in use from 1400 B.C. to A.D. 300. However, the source of gold brought to Jerusalem by King Hiram and King Solomon presents a much more difficult problem. King Hiram's initial contribution of 540 talents and King Solomon's later acquisition of 666 talents, totaling 1,206 talents, would have required production of 34,411 kilograms (1,106,224 ounces)—about 34.4 tonnes, which requires a first-rank mine or district. In addition, as will be shown, such production required intense mining, concentration, smelting, and casting in bars for transport. Is it possible that a mine capable of such production existed in ancient Arabia? Information on ancient and modern mining at Mahd adh Dhahab may help decide this question.

Ancient Mining at Mahd adh Dhahab

Over a long period of time, perhaps tens of millions of years, the minerals and rocks associated with the gold were broken down and carried away, leaving a high-gold crust. This crust had a high density (as much as nineteen times as dense as water), which resulted in its accumulation on the slopes of the jabal as eluvial deposits of incredible richness. Some gold was carried out onto adjacent plains, where it formed placer deposits (Bagdady and others, 1976). The eluvial gold was in finely divided form as crystals, wires, and particles that could be easily separated from the associated residual soil mantle by simple washing or winnowing.

There was also rich ore underground in the upper parts of the veins. The heavy gold and silver, mixed with quartz, feldspar (potassium and sodium aluminum silicates), and chlorite (iron, magnesium hydroxyl silicate), together with sulfides such as pyrite (iron sulfide), chalcopyrite (copper, iron sulfide), galena (lead sulfide), and sphalerite (zinc sulfide), had moved downward along fractures in the veins and adjacent bedrock. All of the minerals would have weathered in situ into white to gray alteration products, except pyrite, which alters into red to brown oxides, and chalcopyrite, which weathers to a green carbonate (malachite) and a green silicate (chrysocolla). Malachite has long been prized for use in ornaments. It was instantly recognizable by ancient prospectors, who believed it to often be indicative of the presence of gold and silver.

The biblical account notes two separate periods of production from Ophir, the first yielding 540 talents (1 Kings 9:14, 26), and the second yielding 666 talents (1 Kings 10:14). In our early work at Mahd adh Dhahab, we tested the placer potential on the slopes around the jabal. It seems likely that the initial 540 talents could have been recovered in a few years by a small force of workers under the supervision of skilled overseers. After the gold was concentrated it could have been carried as nuggets or dust in goatskin pouches or melted into bars for transport by camel. This first phase of mining may have ended when readily available eluvial and placer gold had been exhausted.

The second period of gold mining could have involved removing rich ore downward from the surface within and along the veins. Its recovery required crushing the quartz on anvils, coarse grinding on querns, and fine grinding in trough washers (see fig. 11.5). Possibly a few hundred workers could have been involved in this work.

No determination of the grade of ore mined by ancient miners is possible now, but reasonable estimates can be made from figures furnished by O. R. Dolph (1942). He visited Mahd adh Dhahab in 1939 and described the early phases of SAMS operations. Dolph noted that the tailings and dumps from nearby ancient workings totaled nearly 1 million tonnes. Twitchell (1958b) calculated that 293,848 tonnes of ore were sorted from these tailings and dumps. If tailings containing 0.62 ounce gold per tonne were discarded, then the ore treated initially probably contained as much as an ounce per tonne. For example, with ore containing 1.5 ounces gold per tonne, the required 666 talents (610,902 ounces or 19 tonnes) could have been recovered from 407,268 tonnes of ore, assuming 100 percent recovery. With the crude methods of recovery used and resulting loss of gold, however, it seems more likely that 600,000 tonnes would have been required. The gold at Mahd adh Dhahab is mainly in rich lenses and pockets, so only a fraction of the ore mined need have been crushed, ground, panned, and smelted. The 1 million tonnes of material left on the dumps at the time of Dolph's (1942) visit included some barren quartz and rock as well as fragments of grindstones that had been discarded by SAMS before processing.

The preceding discussion is based on the assumption that gold bullion was the primary product of the mining. In SAMS operations the gold:silver ratio averaged about 0.77:1, yielding a bullion high in silver. The gold:silver ratios varied considerably through time, depending on the level from which the ore was being mined. In 1939–1946 the ratios ranged from 1.6:1 to 0.75:1; thereafter, as the proportion of deep underground ores increased, the ratio decreased from 1.27:1 in 1947 to 0.53:1 in 1954 (Luce and others, 1976). If the bullion produced in Solomon's time was also high in silver, then the tonnage and grade required would be much less than the amounts estimated earlier. The gold:silver ratios probably resembled the ratios in early stages of modern mining at Mahd adh Dhahab and may have been 1:1 or greater. Thus, the bullion may have ranged from 500 to 750 fine; 666 talents of bullion of this range could have been obtained from 300,000 to 400,000 tonnes of ore. It is perhaps significant that 6,000 tonnes mined in 1981 for metallurgical tests by Goldfields of London averaged about 1.5 ounces gold to the tonne. It seems reasonable to assume that the ore mined in King Solomon's time was comparable in grade.

A comment on the statement in 1 Kings 10 that "the weight of gold that came to Solomon in one year was six hundred and sixty six talents" is warranted. It is possible that this total may represent the accumulated production of many years, which was then delivered in one year. Judging from figures on production of ancient mines by R. J. Forbes (1963) and C. DeKalb (1914), no mining crew in Arabia or its surrounding regions was capable of producing this quantity (19.0 tonnes of gold from ore) in a year using the technology available in 950 B.C.! It took the Saudi Arabian Mining Syndicate (SAMS), with 700 skilled miners and mill men and a mill capable of treating more than 150 tonnes daily, 15 years to recover 23.8 tonnes of gold.

Dates of Workings at Mahd adh Dhahab

No archaeological work has been done at Mahd adh Dhahab, so there is no direct confirmation that operations were carried on there during King Solomon's reign. However, the record from Lagash, Mesopotamia (Lutz, 1924), indicates that gold was recovered about 2100 B.C. from the drainage of Wadi Ar Rummah, the headwaters of which are near Mahd adh Dhahab.

Further support of early metal mining in the Middle East was furnished by William Overstreet and others (1983) who studied copper, iron ore, and slags from Jordan that date from 1300 to 800 B.C. and include the time that Solomon might have obtained gold from Ophir.

Gary Rollefson reported in a letter written in September 2000 that copper mining in southern Jordan took place as early as 4000 B.C. Malachite is found in surface workings, and it seems likely that local Bedouins were familiar with oxidized copper minerals. Slags and casting molds from copper mines in southern Jordan have been intensively studied recently by T. E. Levy and others (1999), who report

Iron Age (1200–586 B.C.) mining operations during the time of King Solomon's reign.

With metal mining going on in Jordan at that time, it is quite likely that people traveling along the incense route would have noted the quartz veins at Mahd adh Dhahab, would have examined the veins, and would have found the abundant copper minerals in them, as well as a very rich blend of gold and silver crust in the soil. They might then have gathered samples and taken them to Jordan, where knowledgeable metal workers would have recognized copper, gold, and silver.

Modern Mining at Mahd adh Dhahab

The first item on Twitchell's agenda was to determine the potential at Mahd adh Dhahab. The workings were therefore thoroughly sampled in late 1932 (Dolph, 1942; Twitchell, 1958a, b). Preliminary examination indicated that about a million tonnes of tailings and dumps from earlier operations were available. These assayed 0.62 ounce of gold (19.2 grams) per tonne. In addition, the ancient workings contained much gold-bearing rock, forming a basis for an underground operation.

Twitchell tried without success to interest American firms in a mining venture in Saudi Arabia, but after gold prices increased from $20.67 to $35 an ounce in May 1934, the British SAMS was formed to evaluate a concession covering 150,000 square miles in Saudi Arabia. This concession included many ancient mines. During the next year, fifty-five mines were examined, and seven were explored in detail. Of these, only Mahd adh Dhahab seemed to have sufficient reserves to support a significant mining and milling operation. Plans were made to put Mahd adh Dhahab into operation.

Beginning in 1935, mining equipment was moved into the Mahd adh Dhahab area; milling operations began June 1, 1939, and continued until July 22, 1954. In all, 885,048 tonnes of ore, yielding on the average 0.85 ounce of gold and 1.1 ounces of silver to the tonne, were milled by SAMS. Twitchell (1958a, b) states that the ore was partly derived from underground workings (about 591,200 tonnes) and partly from ancient dumps (about 293,848 tonnes); the ore from underground averaged 1.09 ounces of gold; the dumps that had been discarded by the ancient miners averaged 0.62 ounce of gold to the tonne. The total recorded production was 765,768 ounces fine gold and 1,002,029 ounces silver (Goldsmith, 1971; Goldsmith and Kouther, 1971); this aggregated 1,767,797 ounces of gold-silver bullion. Thus the modern production of 23.8 tonnes was derived from sorting through dumps left by ancient miners and by extending the mines to deeper levels. It represents a gold production approximately two-thirds of that earlier obtained by King Solomon.

Summary

At the outset, there were three requirements for identifying a mine as Ophir: (1) adequate size—the district must contain mines of significant size, capable of yielding over a million ounces of gold and silver in ancient times; (2) the mines must lie within a reasonable distance of the Gulf of Aqaba; and (3) the mines must have been productive during King Solomon's time.

Mahd adh Dhahab is of adequate size. Modern production from 1939 to 1954 of 765,768 ounces of gold from nearly 900,000 tonnes of ore signifies a first-rank district. In ancient times a great deal of gold was easily available, initially on the surface, and later in shallow workings in the oxidized zone. Production from surface and shallow underground workings could have reasonably reached the roughly 34 tonnes (over a million ounces) attributed to Ophir in the biblical account. No other mine in the Arabian Peninsula or in surrounding regions approached this production figure, though many mines were worked on a small scale and probably swelled the regional totals.

The mine is also a reasonable distance from the Gulf of Aqaba—travel by land from the Mediterranean to Mahd adh Dhahab would have been possible along the incense route (Al-Woihabi, 1973; Van Beek, 1969). The distance is approximately 525 miles. It would also have been possible to travel by sea: from Eloth to the mouth of the

Gulf of Aqaba it is 99 miles; it is 372 miles farther to a point on the Red Sea nearest Mahd adh Dhahab, and about 150 miles overland from the coast of the Red Sea—a total distance of 620 miles.

Production during King Solomon's time has not yet been confirmed. If mining during that time could be established, my case would be strengthened. Production of gold from the headwaters of Wadi Ar Rummah in 2100 B.C. is recorded in clay tablets of Lagash, and it seems quite likely that Mahd adh Dhahab could have been discovered in the interval 2100 to 950 B.C.; thus, it could well have been the principal source of King Solomon's gold. In addition, evidence from sites near the incense route in Jordan (Overstreet, 1983; Rollefson, 2000) indicate that Mahd adh Dhahab could have been active during King Solomon's time and could have furnished the gold required. Perhaps excavations of old buildings just north of Jabal Mahd adh Dhahab might reveal clues to the dates of early mining in this area. When archaeological investigations are carried out at Mahd adh Dhahab, it is likely that production during King Solomon's time will be confirmed by carbon-dating methods and study of artifacts.

What Became of King Solomon's Gold?

This question was answered by Kitchen (1989), who translated Egyptian hieroglyphs that recorded shipment of 384 tonnes of gold and silver by Pharaoh Shoshenq I (called Shishak in the Bible) to Egypt five years after Solomon's death. The gold and silver were used in offerings to the gods. I hope that some of the gold can eventually be found and analyzed for characteristic trace elements and, thus, compared with gold taken more recently from Mahd adh Dhahab. Such research would allow us to trace the movement of ancient gold throughout the Middle East.

The preceding account involves some speculation, but it provides a basis for testing of my ideas: archaeological work can determine the dates of operations at the site, gold from the site can be studied with neutron activation techniques to determine whether gold objects presently available in the Middle East might have come from it.

While the Bible is not a historical document, it does contain many insights into the life of the ancient Middle East. I therefore regard the biblical story of Ophir with an open mind, while suggesting that the account be tested. If my thesis is substantiated, it is my hope that the name Ophir shall gain its rightful place on the mineral map (and in the mining history) of Saudi Arabia.

Chapter 12

Dreams Turned to Ashes

1960–2000

> The work goes on, the cause endures, the hope still lives, and the dreams never die.
>
> —EDWARD KENNEDY, 1980 speech

> All this glory [turned] unto ashes . . .
>
> —FULKE GREVILLE

Building Our Dreams

We were happy in California through the 1950s and 1960s. We built a small apartment house near the business district of Palo Alto; the boys finished high school and went off to college. Michael and Steven married in the late 1960s and continued their studies in graduate school, finishing their doctorates in the early 1970s. Kim was studying electronics at California Polytechnic University. I wrote a "letter to three sons," congratulating them on their accomplishments. Arleda and I had taken up our alternative lifestyle in Arabia in the mid-1970s, and the future looked rosy. Then things started to unravel. Arleda began to show signs of memory loss, and in 1977, fate struck: Steven was drowned in an accident. But let's first look back at happier days. . . .

FIG. 12.1. *The Roberts family: seated, author and wife, Arleda; standing (left to right), Michael, Steven, and Kim, ca. 1956.*

FIG. 12.2. *The musicians: Steven with French horn and Michael with trombone, 1963.*

MICHAEL

From the time the boys were very young, I had tried to instill the idea that science could be fun and challenging. Maybe I was following the pattern of my own dad. Michael says that by the age of thirteen, his interest in biology had begun to blossom. He loved to hike and watch birds, and his trips got him to thinking about life's diversity. His early practical experience with animals was tied to the large gopher population around our house in Menlo Park. We had a Siamese cat that liked to catch gophers, and she would deposit them neatly on our back step. Mike delighted in skinning and stuffing the gophers. He amassed a large collection of skins, which he later donated to the University of California's Museum of Vertebrate Zoology.

He obtained his bachelor's degree in zoology from the University of California at Berkeley in 1966. One of his favorite courses was a class in vertebrate paleontology, which gave him an opportunity to join an archaeological dig in Wyoming—the Hell Gap Site run by Harvard University. He spent two summers there, and when they got to layers formed in late Pleistocene times, his knowledge of small animal bones proved invaluable in helping to determine what climatic changes had occurred. This work formed his master's thesis at the University of Wisconsin and his first publication.

His Ph.D. dissertation dealt with the thermoregulatory mechanisms of small rodents. With his doctorate in hand, he took a postdoctoral position at New Haven, Connecticut, where he worked at the Pierce Foundation Laboratory in the Yale Medical School on thermoregulatory control of blood flow. In 1981 he accepted a position in the biology department of Linfield College at McMinnville, Oregon, and he is currently a professor of biology, teaching physiology, genetics, developmental biology, and human evolution. He directs a research grant in the biology department and is continuing his research. He and his wife, Sherill, a professional musician, have two daughters, Rosemary and Amelia.

STEVEN

One of my poignant memories of Steven is of an incident when he was in the sixth grade in Menlo Park School. He had been asked to write a poem about his father, and he wrote,

My Father

My father's hair is black as night
His eyes are brown and shrill
His expression shows the mood he's in
If it's happy, sad, or still.

An outdoor man he is, my Dad
He knows how to take it all
Be it wind or rain or sleet or snow,
He'll fight it through spring and fall.

When the weather's good he loves the outside;
When its bad he just stays inside
And thinks about times he will have
When the rain doesn't make him hide.

He's healthy and happy and sternly set
With magnificent teeth and bone
Thru all, including thick and thin
My father, he has shone.

(Chico Roberts, February 29, 1960)

Needless to say, I was deeply moved, and I have treasured this heartfelt poem.

Steven finally settled on geology as a major course in his junior year. I tried not to deliberately influence any of the boys to follow in my footsteps, but a childhood filled with USGS trips to and discussions about Nevada and Utah had ignited a spark in Steven. When he told me he wanted to go into geology, I could not hide my delight with his choice. I was extremely happy when he informed me that Charles Meyer, his professor of economic geology at Berkeley and a man whose work I admired, had invited him to work summers at Butte, Montana, for the Anaconda Copper Company on geology of the Contention and other mines. After graduation, he was accepted into the Ph.D. program at Harvard. In 1996, Ulrich Peterson, a profes-

FIG. 12.3. *Steven at Harvard University holding a model of a tetrahedrite crystal, 1976.*

sor at Harvard, told me that Steven had been his most competent student in economic geology, a comment I will always treasure.

Steve and his wife, Dinah, moved to Cambridge, Massachusetts. After finishing his dissertation, Steve returned with Dinah to Montana to rejoin the geologists at Butte. One day in the spring of 1977, he and some friends decided to kayak down the dangerous rapids of the Madison River in central Montana. Steve had a two-person kayak, and the rapids demanded the strength and skill of two men, but Steve said if it was so dangerous, why risk two lives? He took the kayak out on his own. He was navigating just fine until a very strong current caught the kayak. He was thrown out and then was sucked under a large boulder. His two companions, forty feet away, could only watch in horror as it happened, afraid to make any attempt to save him. He was only thirty years old. The family all met in Butte, and we began to try to figure out how to deal with the enormous loss of son, brother, and husband.

Steven didn't have the chance to fulfill his great potential, but he did make several positive contributions at Butte. He studied the igneous rocks that were genetically related to the ores, recognizing that the magmatic quartz in the granitic rocks contained fluid inclusions that revealed much about the geochemistry and history of the late magmatic phases as well as the early history of the hydrothermal stage. He was told that the inclusions, which contained small amounts of liquids and gases, could not be used effectively. Steven proved that they could.

Once, while Steven was there, a group of USGS economic geologists and college professors toured the Butte mine. Harold Helgeson of UC Berkeley told of the early history of the granitic rocks, pointing out that corundum was one of the early minerals to form, but that it commonly was resorbed during later magmatic history. After Helgeson's talk, Steven was asked to present the results of his work. Steven began, "Dr. Helgeson, corundum is indeed an early mineral at Butte, and I have identified it in thin sections." Then he flashed a photomicrograph on the screen with a crystal of corundum in the center of the slide to clinch his statement.

At Butte, Steven recognized in the deep levels of the underground mines the characteristic minerals of the peripheral zones of disseminated copper metallization and predicted that another ore body was present at depth in this area. Deep drilling proved this prediction true, and Steven was credited with the discovery of the new deep ore body. I saw in him my own passion for research and discovery, and it allowed him to leave behind a legacy.

The Anaconda Company has gone out of the business of mining, but other companies are carrying on exploratory work at Butte. At a meeting of the Society of Economic Geologists in November 1999, three of the participants singled out Steven's work as an important contribution to the early metallization stage of porphyry copper deposits.

In 1979, Arleda and I established an endowed scholarship at Harvard University in memory of Steven. Many of his colleagues at Butte, USGS geologists, family, and friends contributed. In fall 1999, John Dilles of the University of Oregon (letter, December 2000) told me that the Society of Economic Geology in November 2000 sponsored a symposium on research entitled "New Insights Into the Giant Butte Hydrothermal Deposit." He went on to say, "As you know,

[Steven's] Ph.D. work on the Butte District [at Harvard] in the early 1970s provided some of the most important advances in the understanding of the Butte geology and porphyry copper deposits in general. Our research over the last five years has built on Steve's studies, and we have found that his work was both accurate and much ahead of its time. In the symposium I wrote one abstract with Steve as coauthor because of the importance of his contributions. Mark Reed's abstract and presentations also made use of Steve's work, and Ph.D. student at U. of Oregon, Brian Rusk, has confirmed many of the initial results Steve reported on fluid inclusions he undertook at Harvard thirty years ago. We plan to publish a series of papers on the Butte district over the next few years, perhaps as a special volume of Economic Geology. At least one paper will be coauthored by Steve Roberts, and will contain some of the work he reported in his Ph.D. thesis."

Arleda and I directed that the scholarship funds be used for deserving students studying economic geology. After the economic geology program was eliminated at Harvard, the scholarship was transferred in 1996 to the Mackay School of Mines, University of Nevada, where it assists deserving students in research in economic geology and is used to help administer the Ralph J. Roberts Center for Research in Economic Geology.

Steven's untimely death was devastating. Arleda and I mourned him even while the gap in our lives was largely filled by Michael and Kim. His ashes will rest with ours in the cemetery at Rosalia, Washington. *Masalama* Steven.

KIM

In school, Kim played shortstop on one of Menlo Park's Little League teams. He also loved golf and softball. He majored in physics and chemistry at the University of California at Berkeley for the first three years, then he dropped out of school for several quarters, remaining in Berkeley as manager of a commune. He learned the facts of life very quickly at the commune, as the manager did most of the work at the house.

While Arleda and I were in Saudi Arabia during the 1970s, Kim returned to school, this time California Polytechnic. He took electrical engineering and finally found his niche. Because of the huge demand for electronic engineers in Silicon Valley, he left school just before graduation to work for a firm in San Jose, California. Things went very well for the next few years, but the company eventually failed, and Kim was cast adrift. He then joined another firm, where he worked as a consultant, writing programs for various processes. In 1990 he moved to Reno, Nevada, to work with an engineering firm as a computer specialist and is in 2002 finishing his B.S. in geology at the University of Nevada's Mackay School of Mines.

Kim came to my assistance during the late 1980s when Raul Madrid and I bought the Alvarado gold mine in Arizona, twenty miles northeast of Wickenburg. We tried to put the mine into production, but the ore near the surface was too low in grade. Kim then took over the project, finding a market for decorative granite boulders that bordered the vein. The operation was a far cry from gold mining, but it was successful for a time; in 1998 we ceased operations. Kim's business skills and willingness to assist me in my business and personal affairs have given me tremendous comfort and peace of mind.

Losing Arleda

Michael, Kim, and I will always remember Arleda as sunny, bright, very active, and always enthusiastic about life. Seeing her rapid mental decline was devastating. She had always been involved in a flurry of activity with multiple projects going around the house and the community. One notable event was the Philippine Festival she organized in Palo Alto. With help of Philippine teachers, she put on a brilliant evening of music, food, and entertainment. In 1963, Michael and Steven were in the California Youth Symphony, and the group had a chance to go to Japan (see fig. 12.2). No American youth orchestra had ever performed outside the United States before, and Arleda was determined to seize this opportunity. She rallied her volunteer troops to raise money for the trip. She even talked her way into the office of the CEO of Levi Strauss and won that company's support. Her success in these situations demonstrated her unusual ability to tackle a range of projects and see them through.

Arleda started showing signs of an Alzheimer's-like condition in 1975, but they were subtle and it took me a while to realize what was happening. I think that Steven's death caused a deep depression that hastened her decline. At first, I began to notice that whenever we were around friends, she would quietly ask me to repeat their names. Then, after I told her, she would go up to them and greet them graciously.

I had to return to Saudi Arabia in 1982, so I took her back with me. We stayed in a USGS staff house, where a cook and houseboy could take care of her simple requests. We returned to the States the next year and I started working with VEKA, so we moved from Palo Alto to Winnemucca, Nevada.

I realized that I could not continue my work and take care of Arleda. My sister, Louise, who lived in Oregon, told me that she knew of a nursing home with a family atmosphere. I felt that this would be a better environment for Arleda, and this proved to be the case. My mother joined her a few months later. Michael and his family lived nearby and were able to visit Arleda frequently; Kim also visited whenever he could. Arleda died July 6, 1988.

My Fading Vision

Over the years I have been blessed with general good health. Sure, I have had my share of problems, but I figure that they come with the territory.

By far, my worst problem during recent years has been macular degeneration (i.e., the maculae, the areas on the retinas where images are brought into sharp focus, have begun bleeding). This was first diagnosed during an eye exam in the 1960s. The doctor noted definite signs of deterioration, but he told me no treatment was available at that time, and he said that the disease would develop slowly. I didn't have real problems until the late 1980s.

During a routine eye examination in 1988, another ophthalmologist noted signs of capillary bleeding, which had distorted the retina. I was referred to a retinal specialist, who said that a laser could be used to cauterize the capillaries and stop their bleeding.

That treatment stopped the bleeding, but by 1994 I could no longer read normal book or newspaper type. I found that I had to use ever larger type in my Macintosh computer; now I read letters nearly an inch high. Thanks be to the inventors of the Mac. Now I can also read regular print when it is enlarged—with a magnifier—to an inch high. The retinal specialists say that I will probably not lose my eyesight altogether, but I notice a gradual decrease in my visual acuity and my peripheral vision. Still, I am luckier than many of my friends who have passed on, so I count my blessings each day. I give great credit for my overall good health to my genetic good fortune—I had long-lived grandparents and parents—and to many years of strenuous and invigorating exercise, hiking the mountain ranges of the West.

Chapter 13

VEKA

The more original a discovery,
the more obvious it seems afterwards.

—ARTHUR KOESTLER

After forty-four years and eleven months with the USGS I finally retired in 1981. I was only seventy years old and was certainly not ready to hang it all up. I wanted and needed to be productive. I had been planning my move to industry for some time. I had completed final reports on my work in Nevada and in Saudi Arabia. So I began looking around for something to do. But I was a little shy about selling my services to mining companies, something I had never done. I attended professional meetings, where I dropped subtle hints to geologists working for companies with active exploration programs. I received plenty of encouragement but not much else. One geologist knew I had a fair retirement income, and he advised me not to work too cheaply, because that might affect how much other consultants were able to charge.

This sad state of affairs continued for nearly a year. Then Robert G. Reeves, a former USGS colleague, and I attended an AAPG meeting in San Francisco in 1981.

Bob was then teaching at the University of Texas of the Permian Basin. Knowing that I was on the loose, he suggested that he, Victor E. Kral, and I form a consulting firm and pursue our own projects. That sounded like an excellent idea to me, and I was eager to meet and make firm plans.

FIG. 13.1. *Robert G. Reeves.*

FIG. 13.2. *Vic Kral and author at author's eightieth birthday party, 1991.*

We held our first meeting at Vic's home in Sparks, Nevada, and agreed to form a consulting business. Later in Odessa, Texas, at the Reeves's home, we discussed several options, but none worked. We finally decided to form a company that we would finance ourselves. Our first objective was to pick up properties in certain critical areas in Nevada where there might be a hope of finding an ore deposit. We named our company Victor E. Kral Associates (VEKA) after Vic. Vic was widely known in the mining industry and had an excellent reputation as a mining geologist in Nevada. He was the perfect choice for our first president.

Bob, Vic, and I had worked together in the early war years. Vic had left the USGS to work first with a small exploration group, then with the Nevada Bureau of Mines, and later with the Ford Motor Company and the Hanna Mining Company. Vic and Bob had both graduated from Nevada's Mackay School of Mines. Vic had earned a degree in mining engineering, while Bob had earned a degree in geology and gone on to get a Ph.D. in geophysics from Stanford University. Later, another Mackay graduate, Gerald Hartley, joined the group. Like Vic, Gerry was a mining engineer. Together we made a strong team. Gerry and Vic had an engineering approach, and Bob and I had the geologic approach. Drawing on my background in Nevada's regional geology, I quickly suggested several possible projects. I proposed that we first carry out a staking program in likely areas within the Battle Mountain, Carlin, and Getchell mineral belts, searching for ore bodies that might continue down dip or along strike from known ore bodies. We also looked for areas where potentially favorable host rocks might be found in likely structural settings, as close to the surface as possible.

During the first year (1982) we staked about 1,250 claims and then began to do a little exploratory work. We had soon run through our available capital, and so we began to look around for partners who might want to invest in our program. The response at first was a deafening silence. One geologist, Allen Park, visited our properties and wrote me after the visit. He said, "I think your properties have about one chance in ten thousand of a significant discovery!" I took this as a challenge. Since we had only about fifteen groups of claims, the odds were not good.

I had put most of my terminal pay from the USGS into our project as well as a year's worth of consulting income. Our fifteen blocks of claims were scattered throughout northern Nevada. We knew they were not sure bets, but we did believe they had a good chance for paying off. First, however, we badly needed cash. We offered our claims to many individuals and companies without success. What were we to do? Fixed expenses, such as assessment costs and exploratory work, quickly depleted our available cash. Where to turn?

Fortunately, Vic found a way to save us. He had on occasion consulted for Andrus Resources, a petroleum company in Houston, Texas. Andrus was primarily in oil and gas production, but it made occasional forays into the metals field. We had been turned down by many of my old "friends" in and out of major corporations. Vic proposed to Andrus Resources that it finance our exploration projects, and surprisingly, the people in charge agreed. We ended up with a 50/50 partnership with the company. This was a windfall, and we were delighted. Most companies had offered partnerships in which they would receive 60 to 70 percent of any profits, and we would receive 40 to 30 percent. Bill Andrus and his associate, Charles "Butch" Robinson, seemed like they would be good partners. Best of all, Andrus Resources gave us full responsibility for the geologic program, a decision they have never had cause to regret.

We began exploration in the southern part of Battle Mountain at the Modoc Mine. It contained a small gold-bearing vein system, and we believed there was potential for a gold deposit in depth. We drilled and found the potential ore zone to be too deep to be profitably operated at that time, so we turned to a project at Marigold.

The Marigold Mine

The Marigold Mine had been operated on a small scale during the 1930s and had yielded a few thousand tons of gold ore containing about 0.25 ounce of gold per ton. The area had been explored by many companies after mining ceased in the 1940s. The ore body had been drilled down dip very thoroughly with mediocre results. Exploration had been carried on in the footwall zone of the deposit, also with poor

FIG. 13.3. *View of Marigold ore body in Antler Sequence, underlying the Golconda thrust plate.*

results. Our project was to explore the property north along the strike out into the valley, where the continuation of the ore body would be entirely covered by alluvium. Most geologists involved in exploration are leery of such projects, because the alluvium can be hundreds to thousands of feet thick.

Because there was no obvious major range-front fault, I thought that the target would be shallow, and fortunately, I was right. We hit mineralized rock in the first drill hole, encouraging us to move south to a second hole, which yielded fair gold values. The third drill hole encountered material low in gold and silver but higher in base metals. I considered this a discovery. At this time we were in touch with many mining companies who were monitoring our exploratory programs. One of these companies was Cordilleran Explorations (Cordex), a division of RayRock Mining Company. Andy Wallace, Cordex's director of exploration, offered us a deal on the property based on our promising drill hole data. We accepted, and Andy continued the exploratory work.

By the time the drilling was completed, Cordex had found gold reserves of about 350,000 ounces in an ore body that could be mined in an open pit. The mining of this ore began in 1989 with the ore treated in a three hundred–ton per day cyanide mill. Low-grade ore

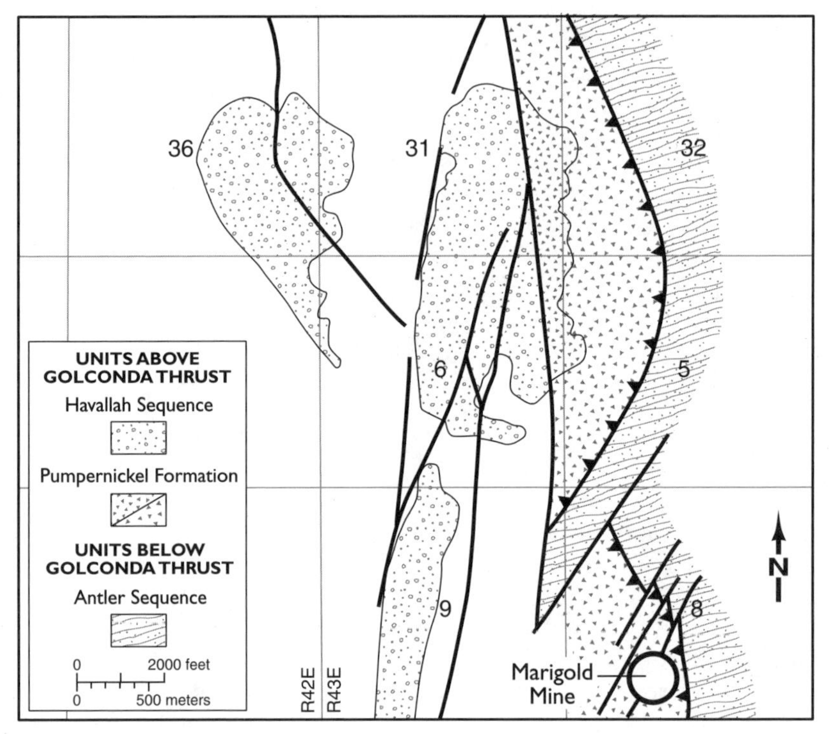

MAP 13.1. *Subsurface geology, Marigold Mine area. The principal ore body is in the western part of section eight (lower right). A new ore body was discovered in late 1999 in section thirty-two (upper right).*

was treated on leach pads. The operation continued until 1995, when the main ore body in VEKA's section was exhausted. In all, about 343,000 ounces of gold valued at about $126 million were produced. There is, however, potential for other ore bodies on the property in zones beneath the main pit and in other areas along strike from the main pit. Low gold prices from 1997 to 2000 have discouraged exploration.

The REN Mine

Another claim group we staked was the REN property in the Carlin Belt. It adjoins the Bootstrap Mine of the Newmont Mining Company on the east, and the Meikle deposit of American Barrick on the north.

FIG. 13.4. *Author and gold bar,* REN *Mine.*

Initially, our strategy was to stake claims that were as close to producing properties as possible. But in this instance Vic was examining the land records in Eureka and Elko Counties to determine the extent of Santa Fe land grant (originally Central Pacific Railroad Company) properties. As most miners in Nevada know, the railroad was granted alternate sections, the odd sections, in a belt twenty miles wide on each side of the railroad right-of-way. Evidently, everybody who had looked at the land position in this part of the Carlin Belt thought that section one (T. 36 N., R. 49 E.) was part of the railroad grant, and therefore could not be staked. Vic enlisted the help of a student at the University of Nevada, who found that section one was outside of the railroad grant and therefore was stakeable. When Vic informed me, I said, "Grab it!" He did, and we found ourselves with a very nice block of eighty-eight claims (also including adjacent sections) in an area of excellent potential. This was a real coup! We named the property the REN block after the topographic quadrangle in which it occurs, the

Santa Renia Fields quadrangle. By 1993, the REN block had yielded a small production of about sixteen thousand ounces of gold for the Cordex Exploration Company. Cordex subsequently subleased the property to other companies, and some excellent gold intercepts were discovered by the Homestake Mining Company, the Uran-Erz Company, and most recently Cameco, U.S. Incorporated, Reno, Nevada. Exploration by Cameco is continuing in 2002.

VEKA has dropped a number of claims from its initial 1,250, but the rest are now leased to various mining companies. If a bonanza gold ore body is discovered on one of our claim blocks, we won't object in the least. So once again, I was without much to do, and I began to tell people that I would be available for consulting jobs; this time a number of interesting projects came my way.

Chapter 14

Consulting

1986–1994

Bell, book and candle shall not drive me back.
When gold and silver becks me to come on.

—SHAKESPEARE, *King John*

Some geologists use witchcraft, forked sticks, and pendulums to find ore, but I and my associates do it a more sensible way—we determine where ore might fit in the geologic framework, then we explore that area.

When VEKA's claim holdings were leased to major mining companies I was left with time on my hands. So, during the late 1980s, I began to respond to offers of consulting work from a number of companies. I evaluated blocks of ground, mostly in the mineral belts of north-central Nevada.

Coral Gold

One company that sought my help was the Coral Gold Company headed by Louis Wolfin. Lou was a pleasant man to work with, and he generally followed my suggestions for exploration—but one time he failed to do so, and it proved costly to us both. During the 1980s, I advised Lou on the exploration of the Robertson property near Tenabo, Nevada. The property was in the Battle Mountain mineral belt northwest of the Cortez Mine and adjacent to the Gold Acres Mine—an excellent geologic setting. The Gold Acres Mine was in the Roberts

Mountains thrust zone, and the ore consisted of mineralized slivers of carbonate rocks and other rocks in an imbricate zone one hundred feet thick. The ore plunged southeast toward the Cortez Gold Mine. I suggested to Lou that he stake the ground out into Crescent Valley. He agreed that this was a good idea and so instructed his mine superintendent. Unfortunately, the superintendent did not favor the project, and he staked only a few claims; he should have staked several square miles, giving Lou the Pipeline and nearby mines.

When Goldfield announced the Pipeline discovery, Lou, of course, was properly chastened. He called to tell me that I had been right, but I was deeply embarrassed, because I could have told him, *Look, Lou, if you don't want to stake the ground southeast of Gold Acres, let's do it jointly.* Instead, I let it go.

In addition to giving advice on exploratory work, I occasionally accompanied Lou on trips to London and once to Paris to explain the potential of the Coral property to prospective investors. Because these trips were a treat—we stayed at elegant hotels such as the Carlyle in London and the Bristol in Paris—I did not charge him for my services on these trips.

St. George Minerals

In the late 1980s Marion Fisher, an independent prospector and businessman from Battle Mountain, suggested that I examine the Betty O'Neal Mine southeast of town. The Betty O'Neal had yielded a fair amount of high-grade silver ore during the late 1800s and was reactivated in the 1930s as prices of metals rose. The easily accessible ore in the Betty O'Neal had already been mined, but a gold-bearing vein on the property had been exposed near the summit of the Shoshone Range at the Dean Mine, and I thought it offered a better target area. I collected some samples from the exposed veins and had them assayed. The assays contained significant amounts of gold and silver. I next contacted several major mining companies and escorted their geologists to the property, but none were impressed by the project until St. George Minerals came into the picture. St. George's president, Ralph Rooney, leased the property. Rooney and his group began to explore the Dean Mine's veins. It had, like the Betty O'Neal, been productive in the late 1800s and early 1930s, but only the upper levels had been thoroughly explored. We decided to test at depth.

James McGlasson, geologist and manager, supervised this work, bringing to the project expertise in geology, mineralogy, and exploration. Raul Madrid and I were hired by St. George as geologic consultants. This exploratory work proved successful, and several extensions of ore shoots on upper levels were successfully projected downward. In previous operations the ore in the upper part of the vein had been mostly stoped and treated in cyanide or stamp mills. Further exploration revealed a rich ore shoot, containing nearly two ounces of gold to the ton below the lowest previous level. So a crosscut adit was driven at a lower level to intersect a rich spot in the vein. This shoot was mined in 1994–1995, but the ore was in small pods, and dilution of the ore during mining and the high cost of mining made the operation unprofitable.

American Barrick Mining Company

APRIL 1986

In April 1986, I was once again in the right place at the right time. I received a telephone call from Peter Chapman, Reno representative of PanCana Mining Company of Toronto, Canada. Peter offered me a consulting job at their Goldstrike Mine in the Carlin district, owned jointly with Western States Mining Company. I was to evaluate the property that adjoined Newmont's Genesis Mine (formerly the No. 8 Mine) on the north and extended to the Bootstrap Mine. I believed it to be a high-priority target area for gold ore bodies.

The property had been in production on a small scale since 1977. Low-grade gold ore had been mined along fractures and treated on leach pads. When I visited the mine, the reserves were about 600,000 ounces of gold in ore averaging about 0.05 ounce per ton. The question was, what was the property's potential?

I asked Rich Seidel, a geologic consultant who had worked with me on projects for VEKA, to assist me and especially to do a magnetometer survey of certain areas. Geologists from Western States—

Keith Bettles and Charles Sulfrian—had kept the mine mapping and drill hole records up to date. They also showed us a 1:24,000 scale (2,000 feet to an inch) airborne magnetometer survey. It indicated the position of the major intrusive bodies that played a part in localization of the ore bodies.

A geologic reconnaissance of the property revealed that metallization was exposed in a limestone unit in the southeastern part of the property in the No. 9 pit area. In addition, Skarn Hill in the southwestern part of the property showed intriguing potential. The rest of the property, however, showed mostly low-grade mineralization in siliceous rocks, such as chert and shale, that were cut by a series of northwesterly striking fault zones. We believed these mineralized faults represented leakages of mineralizing solutions from below and indicated the strong possibility of ore at depth. We were especially interested in mineralization in a region of the mine called the Purple Vein. We also believed that the limestone in the southeastern area and the skarn in Skarn Hill were metamorphosed parts of the carbonate facies of the lower plate. The Post block, another part of the mine, consisting of chert and shale at the surface, contained low-grade gold ore in chert and shale of the Rodeo Creek Formation.

Rich and I modified the geologic map to include this information. We then carried on a geochemical sampling program to test areas where hydrothermal solutions might have mineralized the rocks at the present surface. Then using these data, Rich carried out detailed magnetometer surveys to help determine drill sites.

We confidently recommended sites for drilling, with some holes to go as deep as two thousand feet. We also took the risk of predicting that a world-class ore deposit, containing at least five million ounces of gold, could exist on the property. This implied that the ore reserve potential might increase by a factor of more than eight times. In July 1986 we presented a very positive report on the property to PanCana and Western States. We then met with officials from PanCana and Western States to explain our positive stance and to further emphasize the need for deep drilling.

Buck Morrow of Western States reluctantly agreed to drill deeper than eight hundred feet below the Post ore body area. Before Rich and I

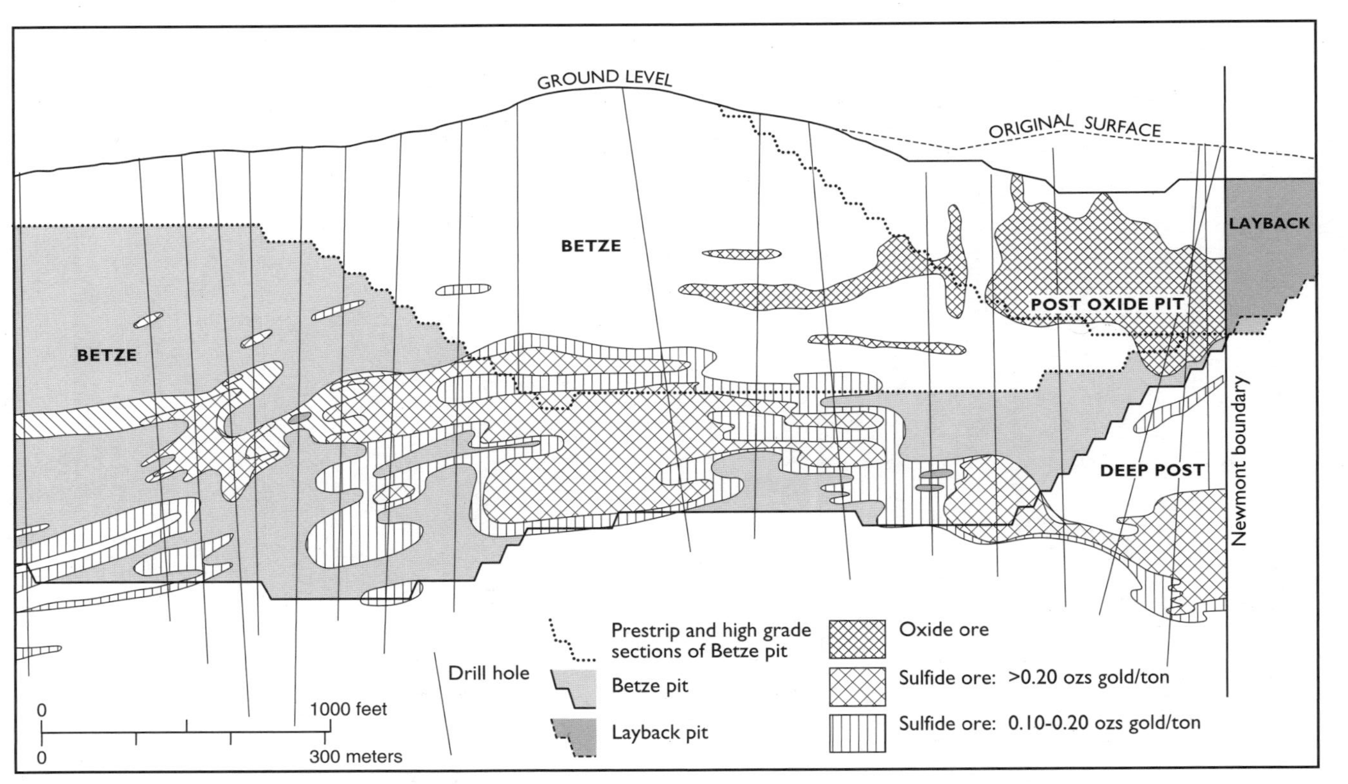

FIG. 14.1. *Cross-section of American Barrick's Betze deposit, Goldstrike Mine. The Post ore body is on the right at the surface; the deep Betze ore body is below and to the left. Areas of high-grade ore are marked with the larger crosshatch pattern.*

left the property, Bettles and Sulfrian reported that they had found a uniformly mineralized limestone at about one thousand feet below the Post area, which contained 0.10 ounce of gold per ton. I announced, "Gentlemen, we are in the lower plate and the prognosis looks great!"

A little later I received a call from PanCana saying that American Barrick wanted to send someone to look at the property with me. It happened that PanCana's president was also on Barrick's board of directors, and our positive report on the ore potential was available to Barrick. Larry Kornze, a geologist at Barrick's Mercur Mine in Utah, was sent to Goldstrike with me. I showed Larry the carbonate facies at the surface in the southeastern area and on Skarn Hill. Then we looked at the drill data for the site below the Post ore body. Larry apparently was convinced; American Barrick is reported to have bought the property for about sixty-two million dollars.

The predicted reserves of at least five million ounces proved to be conservative. Deep drilling revealed a large ore body in the Post-Betze area. By early 2000, Barrick was reporting reserves of more than fifty million ounces (annual reports, 1995–1999).

Coincidentally, during this visit I showed Larry the Purple Vein, about one thousand feet north of the Post ore body. The area had been prospected during World War II for mercury. Although little mercury had been found, I knew that trace amounts of mercury commonly accompany gold, and I suggested drilling in that area for gold. Barrick later drilled and discovered high-grade ore. Larry then brought in Raul Madrid to work out the pattern of the faulting that had displaced segments of the Purple Vein gold ore body. Raul was successful, and the Purple Vein (now named the Meikle deposit in honor of the late Brian Meikle, vice president of American Barrick) is being mined today on a large scale.

In March 2000, Keith Bettles wrote to tell me that recent land exchanges with Newmont have consolidated property, simplified mining operations, and given Barrick more land to explore. Barrick has also started up a new twelve thousand ton per day roaster, which is used to process the carbonaceous ore bodies, especially the western Betze-Post (Screamer), Rodeo, and Griffin deposits. These lie between

Betze-Post and Meikle ore bodies, in the footwall of the Post fault. Rodeo reserves are expected to increase as Barrick continues underground drilling.

Barrick is still exploring and developing Goldstrike. In 2000 Barrick will spend six million dollars on surface drilling, and four million on underground drilling. Recently Barrick has extended the Betze-Post open pit reserves in the Screamer area and delineated reserves by underground drifts south from the Meikle deposit into the Griffin and Rodeo deposits.

Conquistador Gold Ltd.

Raul Madrid received his geology degree from the University of California at Berkeley and took a job with the Bureau of Land Management. Unhappy with the projects available to him, he decided to pursue a Ph.D. at Stanford University, and he asked me to recommend a dissertation area in Nevada. I suggested that he pick a range in north-central Nevada and map it in detail. As Raul proceeded with his project, he discovered that much of the region had been poorly mapped by others, so rather than settling for a single range, he mapped seven or eight of them.

Raul went beyond mapping to study the details of gold metallization in north-central Nevada in ways that had never been done before, using the electron microscope to work out details and new concepts of metallization, and working with coauthors, he published many important papers on the genesis of gold deposits. Having become quite an authority on mining and metallization in Nevada, he completed his thesis in 1987.

Although Raul had an excellent position with the USGS, he felt that he could better test his ideas through consulting work, so he resigned and organized his own company, Conquistador Gold Limited, in Palo Alto, California, in 1987.

His practice did very well the first year. But Raul's enthusiasm led him to take on too many jobs, and he was a little slow in producing some reports, which created tensions with some of his clients.

Raul then suggested that he and I form a partnership. I joined Conquistador, and the two of us worked together well, evaluating prospects and making suggestions for exploration. My forte was looking at the big picture and indicating whether aeromagnetic surveys might be helpful. Raul was the structural specialist, and he made recommendations as to where to look for displaced segments of ore bodies.

Raul's work for American Barrick on its Purple Vein (Meikle) deposit was only slightly short of miraculous; it showed his superb grasp of spatial relationships. Raul was able to visualize ore bodies in space and predict their position in depth when they had been displaced on postmineral faults (faulting after mineralization displaces segments of the vein, making them difficult to follow). He also had a phenomenal memory for details. Even though I had a fairly good memory myself, I was continually amazed by Raul's ability to retain information.

Raul and I jointly evaluated many properties. A description of one of the most interesting follows.

Hi-Desert Gold Company

How naïve can scientists be?
Very naïve. They are so trusting!

—RJR

One consulting job I tackled was proposed by a couple, Sean and Lee Halavais, who were endowed with considerable imagination and intestinal fortitude. They had enjoyed long, colorful careers, and they had found a good deal of gold. The story of their foray into the Carlin Belt is fascinating. They approached me in 1989 and asked me to help them with a project in the area. Their property lay between the Carlin No. 1 and the Goldstrike Mines and included the old Bullion-Monarch (B-M) pit from which several thousand ounces of gold had been produced in the late 1960s.

The Bullion-Monarch ore body was in the right place in the Roberts Mountains Formation, which had been the principal producer in the Carlin No. 1 Mine, but the ore body had been cut off by faults. It could only be traced a few hundred feet to the northwest before it dropped to deeper levels on faults. The B-M Company had tried to follow the ore to the northwest but had found only short segments, so B-M had decided to sell the property. The Halavaises heard about the sale and decided to buy the property. They asked me, as their geologic consultant, to attend a meeting where B-M was going to present technical data concerning the property. I went and listened carefully. I concluded that the property had excellent potential; however, the asking price of $10.5 million seemed excessive. I recommended an offer in the range of $4 to $6 million. The property was in the right place and in the right geologic framework, but it was possible that the targets would be deep.

At that time, I was one of few people recommending deep drilling in the Carlin Belt. The Halavaises decided to take my advice and buy the property, but they paid full price. The deal was closed at $10.5 million. In retrospect, they did the right thing. My strategy might have lost the property to another bidder, though there did not seem to be much competition at that time.

The Halavaises then wanted me to help sell the property to other mining companies. I pointed out that the job ahead was a structural and stratigraphic problem, and that a high-powered structural man like Raul Madrid was needed. The couple dragged their feet for a while but finally agreed.

When Raul and I met with them to discuss the property, they flamboyantly bounced a hundred ounces of gold nuggets from Brazil (worth roughly forty thousand dollars) on the table, and proclaimed, "We are gold finders." They certainly were, and they had a fine intuitive approach to finding gold. Knowing that the B-M pit contained good gold values, we reasoned that the ore body should continue north. They had a good feel for the ore but did not have a sound geologic basis for it. We were able to factor in the geology, which told us that the ore-bearing unit continued north, although it had been displaced on cross-faults to deeper levels.

Armed with this information, the Halavaises set about selling the property. They set up a seminar room in our motel, and we began the hard sell to small groups of geologists and engineers from major mining companies. We had produced appropriate maps and charts, which we displayed to show the property's potential. The Halavaises asked twenty-one million dollars for the property and promised us a percentage of that price if we could sell it. Raul and I agreed and prepared a letter of intent that would have given us part of the proceeds if that price were obtained, but we did not complete an agreement spelling out the details. We proceeded with the "show and tell," taking prospective buyers out to the property, where we described the potential and predicted where a major ore body might be found.

Raul also suggested positions for the first few drill holes. We convinced a lot of people that the project had merit, but none of them wanted to put up twenty-one million. The Halavaises carried on exploratory work that proved Raul's ideas, and ultimately they sold part of the property to the logical candidate, Newmont Gold. Exploratory work apparently went well, and American Barrick made a deal for the rest of it. (Rumor has it that in mid-1996 the reserves were about five million ounces. What a property! Rumor also has it that Newmont paid a significant price for its part of the property, and Barrick is reported to have paid ninety million dollars in stock for its part. Raul and I vowed to spell out our next deal in detail before beginning work on a project.)

Reese River Resources

During the late 1980s, Robert Reeves and I conceived the idea of searching for the southern projections of ore bodies from Copper Canyon into the Reese River Valley. Copper Canyon in the adjacent part of Battle Mountain had yielded more than two million ounces of gold in the Antler Sequence beneath the Golconda thrust plate. Our project seemed sound, but it had one drawback—the targets were likely to be deep.

Where these targeted rocks cropped out on the surface at the Copper Canyon Mine, Fortitude ore body, and other nearby ore bodies, exploration and development were relatively simple. One had merely to follow ore down dip or in fault zones.

In Reese River Valley, however, exploratory drilling revealed that the target, the Antler Sequence (Battle Formation, Antler Peak Limestone, and Edna Mountain Formation), was covered by the Golconda thrust plate. Elsewhere, Tertiary volcanic rocks covered potential ore zones. Locally, Triassic sedimentary rocks are mineralized. We proposed a project to Lou Wolfin and his associates in Vancouver, and they agreed to support exploration. Jon Broderick managed an aggressive exploration project on Reese River, first exploring Triassic rocks near the Echo Bay Mine and later drilling into the Golconda Plate in the northern part of the project area. Although we found traces of gold and silver mineralization, our drills could not reach the targets of Antler sequence rocks, and funds for the project were withdrawn in 1990. I financed a final two drill–hole project myself, but at fifteen thousand dollars a hole, my meager resources were soon exhausted, and the company failed in 1991. It was a good idea whose time had not yet come. Ultimately, metal prices may rise high enough to encourage mining companies again to look again at some targets of the Reese River project. The geology is sound, but the targets are deep.

My Eightieth Birthday

One of the notable events in my later life was a wonderful surprise—my eightieth birthday celebration at the home of Raul and Helen Madrid. Early in the week, Larry Kornze called me and asked me out to lunch. We met according to plan, and over an enjoyable lunch, Larry filled me in on the recent exploratory work at American Barrick's Goldstrike Mine. After lunch, Larry just continued talking rather aimlessly, then he suggested that we go to the Madrids' home to talk to Raul. When we opened the door, there were forty people gathered, singing "Happy Birthday"! I was flabbergasted! Here were the friends and associates with whom I had spent the years since retirement, all gathered together. I was deeply touched, particularly when Raul presented me with a plaque of the Nevada map, on which were engraved *my* mineral belts and some major gold deposits. I was deeply moved by their tribute and recognition of my work.

FIG. 14.2. *Author's eightieth birthday party (left to right): J. McGlasson, Raul Madrid with Elizabeth (partly hidden by map), author (with map of Nevada mineral belts), Lyle Campbell, and John Livermore.*

My Philosophy of Exploration

Philosophy can be defined as "the principles or laws of a field of knowledge." My philosophy of exploration includes knowledge of geologic framework, good land work, and perseverance.

An essential requirement for geologists working in north-central Nevada is that they go to the areas where these major elements of geologic framework were originally worked out, become acquainted with the stratigraphy and structure, then tackle the geologic problems in their local areas. Working only in a local area can be risky, as essential regional elements may be obscured by structural complexity and cover.

The analysis of the geologic framework of Nevada has been assembled over the past one hundred years by giants in geology, beginning with Walcott and Hague during the late 1800s in the Eureka area, and has been bolstered by Kirk, Merriam, Anderson, Ferguson, and a host of others. This framework includes mid-Paleozoic movement of oceanic facies rocks onto the carbonate shelf during the Antler Orogeny and late Permian–early Triassic movement of the Golconda thrust plate onto the Antler Sequence during the Sonoma Orogeny. Mid-Mesozoic folding and faulting and Tertiary faulting are also significant elements of tectonic framework.

Exploration for precious-metal deposits in north-central Nevada has followed a predictable sequence. In the first stage, from the late 1800s through the mid-1960s, only ore bodies that cropped out were discovered. In the second stage, in the late-1960s and through the mid-1980s, when the geologic framework became known and mineral belts were recognized, geologists were able to narrow target zones; most exploration, however, was for shallow step-out targets. Now, exploration for precious-metal deposits has reached a third stage, deep exploration for targets in zones below known deposits and in areas covered by alluvium, volcanic rocks, and thrust plates. The targets are commonly at stratigraphic and structural intersections whose exploration requires geologists with broad and sound regional geologic and technical knowledge to define favorable zones by extrapolating into the unknown.

Chapter 15

Mij

> The most beautiful thing we can experience is the mysterious. It is the source of all true art and science.
>
> —ALBERT EINSTEIN, *What I Believe*

How do two souls unite after sixty-four years apart? It is simply this mysterious universe around us.

In the spring of 1992, I received news that would have a profound effect on my life. I was visiting my sister Margaret in our hometown of Omak, Washington, and she had heard from friends that Marjorie Courtright Ogden, now widowed, was living in Tucson, Arizona, and that her brother, Don Courtright, my old friend, was visiting his sister. I hadn't seen Don since our high school days, when we spent hours huddled over a shortwave radio we had built, listening to transmissions, so I asked her for the telephone number there.

One of Marjorie's daughters answered the phone, and she said, "Don just left for the airport to go back home to Frederick, Maryland, but Mother's here, do you want to talk to her?" I thought *Mother?* Memories swirled in my head. "Mother" was Marjorie, Don's sister, the popular, sweet, smart, attractive brunette whom I had secretly admired. I'd never found the courage to indicate how I felt, nor had I seen any evidence that she was interested in me, although I could remember a high school theater production during which Marjorie had playfully held my hands as we waited offstage for our cues to enter.

In the fall of 1928, Marjorie had enrolled at Whitman College. The following year I had begun my studies at Washington State College.

Our paths hadn't crossed again in more than sixty years. On one of my visits home, I had heard that she was engaged to a young man at Whitman, and in 1932, I learned that they had married.

Now, sixty-four years after we'd held hands backstage at Omak High, my heart began to race at the thought of speaking to her. With some trepidation and a strange, schoolboy nervousness, I said, "Yes, I would like to speak with her." Perhaps Marjorie felt nervous, too, because for some reason, she told me later, she informed her daughter that she would take the call in another room. When she answered, I said, "Hello. Do you remember Ralph Roberts from Omak High School? We graduated in the same class in 1928." She said, "Of course, I remember you." I then chatted, asking meaningless questions, while a feeling of excitement seemed to build inside me. I finally asked if she ever came to the Northwest. "Yes," she said, "I usually spend six months a year in Gig Harbor, Washington." So I said, very casually, "Maybe, I can drop by sometime and see you. I now live in Seattle part time." She mentioned that she would be attending her Whitman class reunion in May but would be back on May 20.

As fate had it, I was needed in Seattle just about then to sell a duplex I owned; I was preparing to move back to Nevada. I dropped in on Marjorie at Gig Harbor. There we were after sixty-four years, practically a lifetime of building and living our own dreams, and learning to survive when some of them shattered. She asked me in and we began to talk, taking turns filling each other in on what we had done in the long years apart. We did not come close to finishing our stories that day. We agreed to get together again, and as I said good-bye and headed to Nevada, I wondered if this strange turn of fate could possibly lead to anything serious!

I found myself seeking excuses to return to Seattle or, more precisely, to Gig Harbor. We saw each other in July and again in August 1992. By then, we'd finished telling each other our stories, but the sparks of mutual attraction had flared into life again after all those years, and I felt no desire to snuff them out. Neither, I was pleased to discover, did she. We had each lived full lives with abundant measures of joy, pain, passion, and reward. From our experiences, we had gained a deep appreciation for life, and we knew how precious and infinitely more enjoyable it is when it's shared with someone. So we dispensed with protocol, polite games, and courtship. We became housemates and soul mates—first in Gig Harbor, then in Reno, and later in Tucson. Marriage didn't seem practical. We each had children with families and saw no way of meeting terms of trusts and the arcane rules of the IRS.

We wondered how all of our children would feel about our willingness to throw convention to the wind. But they were thrilled and comfortable with their parents' "'90s lifestyle." Actually, I believe they were relieved that we had been so fortunate as to find loving companionship at this time in our lives.

We have made a good life. Initially, we split our time between Mij's Gig Harbor two-bedroom apartment and her park-model mobile home in Tucson. The tiny park-model was only four hundred square feet, just the right size for our life in Arizona. We decided in 1993 to take advantage of an opportunity to settle permanently in Tacoma, Washington. Before we became reacquainted, Mij had expressed interest in living at the Franke Tobey Jones Retirement Estates. I thought that I might like such a place, and we went there to scout out a home. We found sixteen beautiful new duplexes on a quiet, friendly cul-de-sac just a mile from Puget Sound with views of Commencement Bay, the Olympics, the Cascades, and Mount Rainier. Travel around the country and the world, and a lively social life, first in Tucson and now in Tacoma, have given us a great deal of pleasure.

Franke Tobey Jones Retirement Center was founded in 1924 to provide a home for lonely, elderly women. Over the years its mission has broadened, and now, as a nonprofit institution, it caters to a much more diverse clientele. It is a wonderful community that provides basic health care and all the services we could possibly need. There is a close camaraderie among duplex renters that reminds me of compound life in Saudi Arabia, and our social calendar stays full.

The duplexes, while small, are more than adequate. Our rather eclectic decor reflects the lives we've successfully combined: Mij has furniture and keepsakes from her years living in Gig Harbor, sailing in Puget Sound, and traveling around the world. I have my Middle Eastern rugs and memorabilia from years in Saudi Arabia and the American West. Though we both moved here and started anew, I felt

FIG. 15.1. *Mij at the tiller in her sailing days, 1960.*

FIG. 15.2. *Author and Mij in Costa Rica with gaily painted oxcart, 1993.*

settled once my maps, personal library, and quite a few rocks had found their niches. Maybe we jumped the gun a little by moving to a retirement home this early, but it is better to be five years too early than a day too late.

One of our greatest joys has been the opportunity to travel together—either to places that hold special significance or to new areas we want to explore. Our first trip was a cruise from Fort Lauderdale, Florida, around Central America, through the Panama Canal, to Costa Rica, and, finally, to Los Angeles. Mij's daughters, Barbara and Dorothy, and their husbands, Gene and Skip, went along to chaperon us, and we all enjoyed the time together.

At the end of 1993, Mij and I flew to Egypt. It was a heady experience, but the highlight of the trip was a visit to Saudi Arabia as guests of Abdulaziz Bagdady, my friend and former colleague, and his wife, Lee. During our week's stay, Ziz drove me to Mahd adh Dhahab, the mine where we had worked together. The mine was still going strong, producing ore that ran nearly an ounce of gold to the tonne, as our drill cores had predicted twenty years earlier.

Jiddah was no longer the quiet town I remembered from the early 1970s. It had become a bustling megalopolis with incredible traffic jams and residents of many nationalities. At the end of our visit, Mij and I returned to Cairo, then flew to Luxor, the jewel of ancient Egypt. We marveled at the ruins of Karnak, the nearby Valley of the Kings, and Thebes, all breathtaking monuments of a vibrant ancient civilization.

We went next to South Africa. Because of political unrest in the city, we stopped only briefly in Johannesburg, the gold mining capital of South Africa, which is known for its industry, its laborers, its unions, and its long history of racial strife. We quickly flew on to Capetown, a gentler community on the coast, and then to the Cape of Good Hope. We appreciated the beautiful city and the good life Afrikaners so obviously enjoyed. I was struck by the feeling that I had been taken back in time to the American West of the late 1940s and early 1950s, when life moved at a relaxed pace, and we felt we had the world on a string.

Our African experience seemed complete when we toured a game preserve near Kruger Park. Seeing what they call the big five—lion, leopard, rhinoceros, buffalo, and elephant—at close range in their natural habitat was breathtaking. By this time we were ready to rest, and we did just that in Sun City, a beautiful oasis where we could choose to play—at golf or gambling—or just laze in the sun. After three days of rest and relaxation, it was time to return to Johannesburg for the trip home to Tucson via Cairo and Los Angeles. Besides new keepsakes we had acquired together, we took with us many shared memories of a wonderful trip.

I had traced my footsteps back to Mahd adh Dhahab, and once there, I had realized that subconsciously I had needed, while I still had some vision, to see once more the special places that had been such integral parts of my geologic discoveries. I also had a great desire to show Mij these places, to help her see and perhaps better understand the reasons for my passion for geology.

In 1994, we signed up, along with Mij's daughter Dorothy, for an Alaskan cruise. The accommodations promised to be far better than those I'd suffered through in 1935, when I had mapped the Alaskan ranges with J. B. Mertie. My view from the ship was limited, but I saw enough of the coast from Seattle to Anchorage to know that fifty-nine years was time enough for the last frontier to have nearly disappeared. Little villages had grown into cities, and much to my surprise, the salmon-canning industry had left Alaska for other sites.

We later joined a travel group on a tour of the Far East. This trip was purely for pleasure with side trips to Bangkok, several ports in Indonesia, and Singapore. In 1996, Dorothy joined us on another voyage, this time to the Mediterranean and Black Sea. We traveled aboard the *Amerikanis,* along with four hundred Germans, some French and Spaniards, and one hundred and fifty Americans and Canadians, to eighteen ports in ten countries. Before it was over, this international blitz of tourists had stormed ports in Belgium, France, Portugal, Spain, Morocco, Italy, Turkey, Russia, and Greece. We were fascinated and exhausted by the kaleidoscope of cultures and people passing before us. There was little time to rest; we kept pushing ourselves because we were seeing places we would never see again.

Back in the States, we recuperated in the Arizona sun. It was refreshing to have no tour guide telling us what to do. We could step out the front door and pick fresh oranges for breakfast or soak up the quiet solitude of the desert. But soon we were rested, and restless again, and when Mij's daughters, Dorothy and Barbara, and Barbara's husband, Gene, invited us to cruise Baja California and then ride by train through Copper Canyon, Mexico, we accepted. After cruising Baja, we boarded a train at Los Mochis and began an incredible ride through some of the most dramatic scenery I had ever seen. No roller coaster ride could compare to ascending the Sierra Madre Occidental, descending to the valley, and going through as many as ninety tunnels. Mij had made the trip many years before with her husband, their travel trailer loaded onto one of the flatbed railcars. As we descended from the summit and made our way to Chihuahua, we passed through a region filled with old mining camps; we could see vivid stains from copper and other minerals.

We treasured these wonderful times together. In March 1997, Mij suffered an attack of angina caused by a blood clot in her anterior descending artery. The Tucson doctors quickly performed an angioplasty and saved her life. She had another attack a few months later, and doctors inserted a stent into the artery to correct the problem. She slowly but surely regained her strength, agility, and energy. By the fall of 1998 she had recovered sufficiently for us to take yet another cruise, again accompanied by Dorothy, around South America. The Gold Museum in Lima, Peru; the Straits of Magellan; and the Iguazu Falls in Brazil were highlights of that trip.

Growing old is a serious and, unfortunately, inevitable business. Those of us who have made it this far deserve to boast. The journey isn't always easy, but it can be very rewarding. I offer this as hope to the younger generation, and, as one who has already lived a lifetime, I offer this advice: Live today in preparation for old age. Pursue your dreams with passion. Follow the Golden Rule. Take care of yourself physically, emotionally, and spiritually. Learn to forgive and be willing to give. Then love will abide and will carry you through.

Sharing personal feelings, revealing my own shortcomings, doesn't come easily to a man of my age and stubborn independence. I'm trying to learn. I often wonder what would have become of me if I hadn't found Mij. For many years I had silenced my pain while watching the ravages of aging destroy my dear, vivacious Arleda. Deep in my heart, I agonized over the reasons for Steven's death. I sought answers. Why hadn't those people tried harder to rescue him? What would he have accomplished as a geologist if he had lived? And for many years, I carried the hurt inflicted by Tom Nolan's unwillingness to recognize my work. I have also struggled with physical disability. My eyes, vital tools for a geologist, are failing. Macular degeneration is a disease that, at least for now, has no cure.

By 1992 I had become a lonely, cranky old man. My friends and sons were concerned for my health and happiness. My sons saw problems ahead. If I had continued down this path, I am not certain that I would have lived to write this book. But miraculously, good fortune brought me Mij. My nearly lifelong passion for geology continues to be a driving force in my daily existence. Although my eyesight is practically gone, in my mind I can still vividly see mountain peaks and cross-sections of great ore deposits beneath their surfaces. I guess I'm still a dreamer. The field of geology is constantly expanding, and I continue to be excited by the new discoveries young scientists make every day.

Chapter 16

Epilogue

The Geological Survey was my life for almost half a century. When I joined the USGS in 1939 it was a small, tight network of dedicated scientists with the freedom to select and carry out unique research projects, so long as they fit into the overall program.

Of course this changed with the strategic mineral program begun by Foster Hewett in the early days of World War II. Foster understood the needs of industry in wartime, and he hired geologists to study and evaluate strategic mineral deposits. After the war, the USGS continued to expand, and new programs were added. But expansion doesn't go on indefinitely; fortunately, I was already gone when a large-scale reduction in force took place in October 1995. More than nine hundred geologists, scientists, and technicians were offered early retirement, demoted, or laid off, and as in industry, the anxiety only increased competition for jobs. Programs continue to be cut back drastically, and the present-day USGS lacks the necessary number of scientists to conduct the basic scientific research our government needs if it is to adequately study, conserve, and use our natural resources.

I realize that change is inevitable, but I can only worry about the effects of such drastic reductions. So, I have decided to spend my remaining years helping to educate the geologists who will be the discoverers in the new century. In 1995 I helped set up a lecture series in economic geology at Nevada's Mackay School of Mines. The lectures focus on Carlin-type gold deposits, specifically those in north-central Nevada. Since our discovery of these deposits, it has become clear that they are a type not tied to a specific locality.

Fig. 16.1. *Odin Christensen (left), of Newmont Mining Company, after delivering first* CREG *lecture, and author.*

Fig. 16.2. *Author flanked by Michael (left) and Kim at first* CREG *lecture.*

FIG. 16.3. *Tommy Thompson, director of* CREG.

The tremendous efforts of the Mackay School of Mines, under Dean Jane Long, and leaders of the mining industry created the Ralph J. Roberts Center for Research in Economic Geology (CREG). Under the direction of Tommy Thompson it is now producing many excellent graduates who are carrying on fundamental research. Odin Christensen, chief of exploration at Newmont Mining Company, wrote to me in January 2000 to report that he had attended the Annual Research Review meeting of the Center for Research in Economic Geology. He said that it was an excellent technical meeting at which students presented cutting-edge geologic work. He was most impressed by the fact that through CREG, a group of geologists from industry, the government, and academic fields were all working together and communicating freely.

In order to give further support to the geology program at Reno, I moved the scholarship funds given in memory of Steven from Harvard to the University of Nevada in 1997. My dream is that this program on Carlin-type gold deposits will continue to grow. After all, these deposits may yield a hundred million or more ounces of gold in the next fifty years.

I am not aware of any other class of ore deposits that has received such detailed study as those of the Carlin Belt and its cousins, the

Battle Mountain and Getchell Belts. During April 2002, cumulative gold production from Carlin-trend mines exceeded 50 million troy ounces valued at 20 billion dollars. Direct and indirect services add another 14 billion dollars to Nevada's economy. Annual production today is about 4 million ounces, and the reserves are approximately 57 million ounces. I am proud to be part of this picture. As I look back on my years with the USGS, I feel that I was trusted with some of Nature's secrets. It was easy to develop a passion for gold—if for no other reason than its beauty, in its native crystal form, formed into an objet d'art, or as gleaming liquid from the furnace.

For many years I was strengthened and blessed by the love and partnership of Arleda, my wife of forty-six years, and our three sons, Michael, Steven, and Kim, each unique and special. During all these years Arleda and I took great joy in watching the boys grow and develop their own life interests. Each showed tremendous talents. Both Steven and Kim inherited my love of geology and the passion for discovery of ore bodies.

The Future

Although many of the easily accessible ore deposits in north-central Nevada have been found, there are still deep deposits to be considered. For example, the Copper Canyon, Marigold, and Lone Tree gold deposits are in the Antler Sequence, which rests on the Roberts Mountains thrust plate and overrides early and middle Paleozoic carbonate rocks. Imaginative geologists will, no doubt, recognize other potential target areas. Should you ever embark on a search for gold, I hope that some of my methods will help you. The best advice I can give you is quite simple (and you probably already know it from reading this book): become familiar with the geologic framework so that you can use it effectively. Good Hunting!

Appendix

Background on the Men of the USGS

When a man is dedicated to the search for knowledge,
he may follow his quest down many strange paths.

—LOUIS L'AMOUR, *The Ferguson Rifle*

Henry Gardiner Ferguson

Henry Gardiner Ferguson was a fine human being, a top-notch geologist, and a major influence in my life. When I joined his research party in 1939, I had no idea what an impact he would have on me. While I had done some mapping, I was totally unprepared for the complexities of Nevada's geology. Fergie had a thorough understanding of the area and a gift for passing that knowledge on to beginners like myself.

He was born June 21, 1882, in San Rafael, California. Soon afterward, his father, an Episcopalian minister, moved his family back to New England so he could become a chaplain at St. Paul's School in New Hampshire. Fergie went to Harvard University, where he earned his bachelor's degree in 1904 and his master's degree a year later.

His first job was geologist with Cleveland Cliffs Iron Company in the Upper Peninsula of Michigan. Fergie told me that upon arrival he was asked if he minded a "warm bed." He replied, "Not at all," and then learned that a warm bed was one that had been slept in by someone from the day shift. By then, it was too late.

FIG. 1. *Henry G. Ferguson, ca. 1930.*

Fergie worked there for two years, then spent 1907 to 1910 as an economic geologist with the Philippine Bureau of Science. These assignments sharpened his interest in geology, and he entered Yale University to continue studies in geology in 1910. At Yale, he roomed with Alan M. Bateman. In the summer of 1911 he worked with Adolph Knopf of the USGS in the Helena district, Montana. His first assignment was mapping a contact between granite and limestone. On his map he showed every turn and bend. Adolph looked at the completed work that evening and snorted, "Verisimilitude wriggles." Fergie got the point and smoothed out the wriggles.

Fergie finished his course work at Yale in 1912 and joined the USGS; however, he did not receive his Ph.D. until 1924. Fergie took great delight in telling the story of how he finally got his degree: "Alan Bateman, my old Yale roommate, found out that I had just published a Geological Survey Bulletin on the Manhattan district, Nevada, that year, so he asked me to come to Yale for a visit and 'make sure you bring along a copy of your bulletin.'" I got there and Alan took me to meet several other professors from the Department of Geology. We discussed the bulletin, and at the conclusion of the meeting, Alan said, "Fergie, you have just passed your oral exams for your Ph.D. You have also submitted your thesis, and have ably defended it. Your thesis is accepted, and you are hereby awarded your Ph.D. Please remit a check for your tuition fees for this year to the Yale University bursar."

It is important to note that in the Manhattan bulletin (1924), Fergie was the first to recognize Paleozoic deformation in Nevada. It was a historic interpretation for which he deserves full credit, though he did not name the disturbance. I did that in 1949, in my dissertation on the Antler Orogeny.

Fergie was first and foremost a field geologist. The Great Basin was his bailiwick and he published many reports on mining districts there —Mogollon, New Mexico; Manhattan, Tybo, and Gilbert, Nevada; and Allegheny, California to name a few—culminating in the report on mining districts in Nevada in 1929. Fergie completed many other geologic chores, at times being geologist-in-charge of the Metals Branch. He also published a professional paper on the Allegheny district in California in 1932, where he quantitatively calculated the depth of ore formation. But he passed it off with a quote from Gilbert and Sullivan, "merely corroborative evidence to add verisimilitude to an otherwise bald and unconvincing narrative." Fergie's familiarity with Gilbert and Sullivan made him very useful to the writers of lampooning lyrics for the annual Pick and Hammer show during which the mighty and snobs were "taken down."

Fergie and Si Muller tackled the Hawthorne-Tonopah quadrangles in the mid-1930s before taking on the Sonoma Range quadrangle with me as an assistant in 1939. These were both monumental jobs, and just completing them was a great accomplishment, but more than that Fergie was largely responsible for setting up the geologic framework of Nevada as we know it today—a notable contribution. The four Sonoma Range maps in which Fergie was the principal architect—Winnemucca (1951), Mt. Tobin (1951), Mt. Moses (1951), and Golconda (1952)—were the maps that outlined this geologic framework. My Antler Peak map (1952) in the northeast corner of the Sonoma Range quadrangle clarified events related to the Antler Orogeny.

Fergie wrote very well. He used to say, "I don't sign anything I write because I usually answer letters sent to other people," referring to the taxpayer letters, requests from Congress, and requests from other agencies he answered on behalf of the director and chief geologist. At times while he was in Washington he would take over the running of the Metals Branch. Fergie would refer to these times as "when I was Chiefing," and it was obvious that he relished the job. He even competed with Foster Hewett for the position. The two men were quite different geologists; Fergie was a stratigrapher and structural geologist, and Foster was a specialist in ore deposits and mineralogy. Foster won because he was more effective in presenting the USGS program to the Congressional Interior Committee.

Fergie retired in 1957 following a field accident in which a geologic pick head shattered, and some fragments entered his right eye, severely diminishing his eyesight. I continued working with him after he retired on a AAPG paper, "Paleozoic Rocks in North-Central Nevada." He, I, Preston Hotz, and James Gilluly collaborated on the work, published in 1958. It was the culmination of his career, extending the work he had begun in 1924 on Paleozoic deformation in the Manhattan district, Nevada.

I relished my visits to the Fergusons' home in Washington. Their magnificent three-story brick house sat on two large lots in a neighborhood with a number of embassies. An evening at the Ferguson home was an event. Guests generally arrived around 4:00 P.M. At about 4:30 P.M. tea was served. This continued until about 5:30 P.M., when the cocktail hour began. A delicious dinner would follow, with an appropriate wine, of course. Finally, everyone would retire to the den for strong coffee with (optional) brandy. Fergie and his wife, Alice, had a full-time cook and chauffeur, a luxury to which I could easily have become accustomed, but unfortunately, it never came to pass.

Fergie and Alice also owned a farm called "Hard Bargain" on the Potomac River across from Mount Vernon. The farm's name presumably referred to tough negotiations over its purchase price back in the 1800s. The USGS "family" loved to gather at the farm. Sunday gatherings at Hard Bargain meant work. Fergie or Alice would assign us special chores. My first chore in the fall of 1939 was digging out the remains of an Indian village. Most of the fun things had already been excavated, but I helped locate the old postholes for the stockade that surrounded the village. As the holes were regularly spaced and could be recognized by circular areas of darker soil—due to carbonaceous material from the long-decayed posts—it was not a tough job. After a reasonable amount of work had been completed, we were treated to a late lunch and sent on our way.

I was at Hard Bargain one morning in 1941 when news came over the radio that Pearl Harbor had been bombed. We listened incredulously; none of us could believe Japan had done such a thing. Now, we realized, we were in the war for the long haul. The lives of many of us would be changed forever.

In his later years, Fergie spent more time at Hard Bargain, so even before his death in 1966, my role in completing our joint reports had expanded. These reports were important milestones in the geology of Nevada. Fergie's analysis of the major structural features of this region helped me focus on the Antler Orogeny, and I was fortunate to be the one to define and announce the orogeny to the world.

Donnel Foster Hewett

I first met Donnel Foster Hewett in July 1937 while I was working at Charles Park's camp at Staircase, near Lake Cushman, Washington, just outside Olympic National Park. Manganese was one of Foster's favorite commodities, and he had come to explore the manganese potential of the Olympic Mountains. He asked us to climb up to the Triple Trip Mine to see a small body of jacobsite, bementite, and neoticite. I had neither seen nor heard of these minerals, but Foster knew them well from his studies of manganese deposits in other volcanic environments.

He stayed at the camp only a few days, and though he didn't attempt many of our daily hikes, he was intensely interested in our accounts of the size of the manganese deposits and their accessibility.

Fig. 2. *D. Foster Hewett, ca. 1970.*

While their location in the park made it unlikely the deposits would ever be mined, it was very important to know where they were and how the ore could be retrieved in a national emergency.

Foster was born in 1881 in Irwin, Pennsylvania. His father was a mining engineer, and Foster often accompanied him on field trips, where he developed an early interest in ore deposits and mineralogy, the science of identifying and studying minerals. He went on to study metallurgical engineering at Lehigh University in Bethlehem, Pennsylvania. After graduating in 1902, he returned to Lehigh as an instructor in mineralogy. He became good friends with Joseph Barrell, who had recently received his Ph.D. from Yale. Barrell sparked in Foster a new enthusiasm for geology.

Foster's first job in the minerals industry was with the Pittsburgh Testing Laboratory (later Vanadium Corporation of America) as a mining engineer. He reported on deposits of vanadium (an element used in the hardening of steel) in the United States, Mexico, Canada, and Peru. During one trip to Peru in 1906, Foster heard stories of a mine very rich in vanadium called Minas Ragras. He found the Peruvian owner, who showed him a piece of the ore—pure vanadinite—

but told him the property had already been promised to a German company. Foster agonized for days then returned to the owner and told him he wouldn't leave Peru until he'd seen Minas Ragras.

Realizing how serious Foster must be, the owner allowed him to visit the mine. It contained a bonanza deposit of vanadium ore—far richer and larger than Foster had hoped. When he returned to Lima, he began negotiations to purchase the property for his company. He outbid the German company, and the Vanadium Corporation, with the fantastic resources of the Minas Ragras, dominated the world market for many decades afterward. Foster married Mary Hamilton in 1909. That same year the Vanadium Corporation awarded him a stipend for graduate school. He went to Yale, where he studied under his old friend and mentor Joseph Barrell. He originally planned to be at Yale for a year, but the Vanadium Corporation granted him a second year's stipend, and he completed his Ph.D. That only whetted Foster's appetite for research. The best option for a career devoted to research, it seemed to him, was a job with the USGS. He took and passed the examination for junior geologist and began his employment in June 1911. His contributions were significant from the very beginning. Even later, when he was an administrator, routine chores never seemed to interfere with his prolific and solid scientific work.

In 1924 he began field studies of ore deposits at Goodsprings, Nevada, where he first recognized dolomitization as a hydrothermal alteration product. In 1926 he went to Europe and found similar alterations. Foster was one of the first geologists to show that thrust faulting of both Cenozoic and pre-Cenozoic age took place in southern Nevada.

In the mid-1920s, Foster began to worry that the country's rate of inflation was rising to an unsustainable level. Stock prices skyrocketed. He had invested heavily in the securities market, and he decided to reduce his risk. He sold all his stocks and invested everything in long-term U.S. bonds. When the market crashed in 1929, he was able to ride out the disaster. In the early 1930s, he judiciously bought back into the market. His predictions had saved him from disaster and had, in fact, helped him do very well financially.

In later years, he used his wealth to help many students attend college and made gifts to several colleges, including Stanford, Yale, and Lehigh. He gave my son Michael a bond that he used to support his studies at the University of California.

After the war was over, Foster became seriously ill and stepped down as branch chief of the USGS. He and Mary moved to Pasadena near the California Institute of Technology, where he was able to continue his research in mineralogy. In 1949 he was the first to recognize minerals of the rare earth deposit at Mountain Pass, California. Rare earths are silver-colored metals that are useful in making compounds involved in computer screens and lasers. He identified basnasite, a rare earth mineral, which was present in commercial amounts. This mineral contains the elements that make a television screen glow. The property went into production and still furnishes most of the rare-earth minerals used in America.

In 1951 Foster and Mary moved to Menlo Park, California, where they maintained their reputations as gracious hosts of gatherings of geologists from academia, the USGS, and industry. Foster, a giant of a geologist, would often begin a sort of after-dinner lecture by rising from his chair and saying, "Now I am going to say a few sweet words about . . ." He continued his mineralogical studies, mostly on manganese, until his death in 1971 at the age of ninety. Foster Hewett made a difference in the lives of those he touched. He was a man for all seasons.

Norman J. Silberling

Until the mid-1950s the Sonoma Range project was carried out primarily by Fergie, me, and Si Muller. Whereas Fergie and I took responsibility mainly for the Paleozoic stratigraphy and structure, Si concerned himself principally with the Mesozoic stratigraphy of the region. As Si's increasingly ill health prevented him from maintaining an active role in the project, Norman J. Silberling filled the gap.

Norm had been Si's field assistant in 1948 while an undergraduate at Stanford, and in 1950 he had commenced his graduate thesis studies on pre-Tertiary rocks in western Nevada under Si's tutelage. His graduate studies were interrupted by the Korean War, and in the early 1950s, Norm returned to Stanford and, at about the same time, joined

FIG. 3. *Norman J. Silberling, ca. 1960.*

the USGS at their newly opened regional headquarters in Menlo Park, California. He obtained his Ph.D. in 1957, and when Si retired in 1965, Norm replaced him on the Stanford faculty.

Norm suspects that his first USGS project opportunity—geologic mapping of the Humboldt Range with Robert E. Wallace—resulted from string pulling by Fergie, who had retired by then but remained influential. The extensive Lower Mesozoic exposures of the Humboldt Range have close stratigraphic and structural ties to the old Sonoma Range one-degree quadrangle, which led Norm, Fergie, and me to informally collaborate.

In the late 1950s, Norm and I began writing a paper (which was published in 1962) summarizing Fergie and Si's work in western Nevada and adding our more recent observations and interpretations. Some of this research is now outdated, but we did draw attention to an important and puzzling relationship. Si had attempted to divide the Triassic rocks in the Sonoma Range into a western, offshore facies and an eastern near-shore facies, and he believed that a major post-Triassic thrust fault—the Tobin thrust—separated the two. Fergie and Si tentatively equated the Tobin thrust with the Golconda thrust elsewhere in the old Sonoma Range one-degree quadrangle. The Golconda thrust separates two vastly different Upper Paleozoic facies: the shallow-water Antler Sequence of Pennsylvanian through mid-Permian rocks and the coeval deep oceanic rocks (Fergie and Si's Pumpernickel and Havallah Formations) that form the upper plate of the Golconda thrust. Where the Golconda thrust was first recognized, in the Edna Mountains and at Battle Mountain, Triassic strata are lacking, so Fergie and Si interpreted its age as probably late Mesozoic because of its presumed connection with the post-Triassic Tobin thrust.

In our paper, Norm and I suggested an alternative interpretation. Norm's investigations of Triassic rocks in the region showed they were characterized by relatively rapid changes in facies, making any necessary displacement on the Tobin thrust minor at best. We pointed to the profound unconformity that Fergie and Si had mapped between Lower Triassic rocks and the underlying Upper Paleozoic rocks of the Golconda allochthon in the northern Tobin Range. Whereas Triassic strata here lay relatively flat, the underlying rocks assigned to the Pumpernickel and Havallah Formations were tightly folded and imbricately thrust faulted among themselves. Consequently, we proposed the name Sonoma Orogeny for the deformation represented by this important unconformity, and we suggested that the Golconda thrust dated to before the early Triassic episode of deformation. Both the Sonoma Orogeny and the late Permian to earliest Triassic age of the Golconda thrust have been widely accepted, although the degree to which rocks of the Golconda allochthon may have been transported by Mesozoic thrust faults is still debated.

Relationships attributed to the Sonoma Orogeny played an important role in the formation of ore deposits in the vicinity of Battle Mountain. Because the Golconda allochthon consists mostly of chert and fine-grained pelitic (shaly) rocks, it forms a relatively impermeable cover over the Antler Sequence. Carbonate rocks make up much of the Antler Sequence and are a good host for gold and silver deposits. Thus, like the Roberts Mountains allochthon in the Carlin area, the Golconda allochthon served as a caprock during early Tertiary mineralization of the structurally underlying carbonate rocks. In the vicinity of Battle Mountain, carbonate rocks of the Antler Sequence host the ore deposits of Copper Canyon, Marigold, and Lone Pine mining areas.

Glossary

Note: Definitions reprinted from the 1997 American Geological Institute Glossary of Geology are indicated with (AGI, 1997); those that have been modified are noted.

Allochthon. A mass of rock or fault block that has been moved from its place of origin by tectonic processes, as in a thrust sheet or nappe (AGI, 1997).

Alluvium (alluvial). A general term for clay, silt, sand, gravel, or similar unconsolidated detrital material, deposited during recent geologic time by a stream or other body of running water (AGI, 1997).

Anticline. A fold, generally convex upward, whose core contains the stratigraphically older rocks (AGI, 1997).

Antler Orogeny. A late Paleozoic orogeny, named after Antler Peak at Battle Mountain, Nevada (Roberts, 1949). Lower Paleozoic rocks deformed during this orogeny here are unconformably overlain by Lower Pennsylvanian strata. However, in Eureka County, Nevada, Roberts (1951) and Roberts and others (1965) recognized that the orogeny spanned from late Devonian to Middle Pennsylvanian time. The orogeny consists of two major phases (Roberts and Madrid, 1990): (1) episodic early thrusting, ending in late Mississippian time, of Lower to Middle Paleozoic oceanic chert, shale, quartzite, and volcanic rocks onto coeval shallow-water carbonate rocks of the continental margin on the Roberts Mountains thrust, and (2) subsequent uplift along the Antler Orogenic Belt during latest Mississippian and Pennsylvanian time, causing clastic rocks to be shed into eastern Nevada and western Utah. The orogenic belt extends southward into southern Nevada and northward into the Mackay area, Idaho (Skipp and others, 1979), and into the Colville area, Washington (Bennett, 1937).

Antler Sequence. A sequence of Pennsylvanian and Permian beds that were deposited on rocks deformed during the Antler Orogeny. In the Antler Peak area they consist of the Battle Formation, Antler Peak Limestone, and Edna Mountain Formation.

Autochthon. A body of rock in the footwall of a fault that has not moved substantially from its site of origin (AGI, 1997).

Breccia. Broken rock composed of angular fragments as in a fault zone. Also used for sedimentary units composed of angular fragments.

Carbonate shelf. A broad shallow-water platform on which carbonate rocks were deposited, bounded by a shelf margin and a relatively steep slope into deep water.

Carlin-type gold deposit. A type of low-temperature (150–270 degrees Centigrade) gold deposit characterized by arsenic, antimony, mercury, and thallium as trace elements. The fine-grained gold is in the range of one to thirty microns. It cannot be recovered in a pan but can be analyzed chemically and by fire assay. Carlin-type gold deposits were first recognized near Carlin (Lynn district), Nevada, and defined by Roberts, Ketner, and Radtke (1967). They have also been called "deposits of invisible gold."

Chalcopyrite. Copper, iron sulfide. Weathers to secondary copper minerals, malachite (green), and azurite (blue).

Chert. Rock composed of hard, fine-grained (cryptocrystalline) silica.

Chlorite. An iron, magnesium, hydrated silicate, generally dark green to black.

Chrysocolla. Copper silicate (green).

Craton. A part of the Earth's crust that has attained stability and has been little deformed for a long period of time. Areas are largely Precambrian in age (AGI, 1997).

Detritus. Sedimentary debris or alluvium carried out into a valley or trough.

Dike. A tabular igneous intrusion that cuts across the bedding or foliation of the country rock (AGI, 1997).

Eluvial. Debris derived from rock weathering in place, which has undergone little transport.

Eugeosyncline. A geosyncline in which volcanism is associated with clastic sedimentation; a volcanic part of an orthogeosyncline, located away from the craton (Stille, 1940) (AGI, 1997).

Fanglomerate. A sedimentary rock consisting of waterworn fragments of various sizes deposited in an alluvial fan (AGI, 1997).

Fault. A break in the earth's crust along which one side may move relative to the other side; commonly steep.

Feldspar. Potassium, sodium, or calcium aluminum silicate, weathers to clay.

Fracture. A break in the earth's crust along which there is little or no movement; may be zones of weakness along which magma and mineralizing solutions might rise.

Galena. Lead sulfide, weathers to white lead carbonate or sulfate.

Golconda thrust fault. A major thrust fault commonly associated with the Sonoma Orogeny of late Permian and/or earliest Triassic age, which carried oceanic chert, shale, and volcanic rocks east onto shelf rocks of the Antler Sequence and related rocks (Ferguson, Roberts, and Muller, 1952; Silberling and Roberts, 1962).

Hydrothermal deposit. A mineral deposit formed by precipitation of ore and gangue minerals in fractures, faults, breccia openings, or other open spaces, by replacement or open-space filling, from aqueous fluids (AGI, 1997).

Nappe. A sheetlike, allochthonous rock unit, which has moved on a predominantly horizontal surface. The mechanism may be thrust faulting, recumbent folding, or both (AGI, 1997).

Oceanic Formations. Name proposed for ocean-basin lithostratigraphic units (e.g., chert and shale) (modified from AGI, 1997).

Plate Tectonics. A theory of global tectonics in which the lithosphere is divided into a number of plates that move independently and interact with one another, causing seismic and tectonic activity (Dennis and Atwater, 1974, p. 1031) (AGI, 1997).

Quartz. Crystalline silicon dioxide; occurs mainly as quartz veins, crystals in granitic and volcanic rocks, and as grains in clastic sedimentary rocks.

Roberts Mountains thrust. A thrust fault of regional extent in north-central Nevada that carried oceanic rocks of Early and Middle Paleozoic age east as much as 120 miles onto coeval rocks of the continental shelf. The thrust was first described by Merriam and Anderson (1942), who thought it formed during late Cretaceous or early Tertiary time. Roberts (1949) and Roberts and others (1965) later proved that the thrust was late Mississippian in age and was emplaced during an early phase of the Antler Orogeny.

Sea-floor spreading. A theory that the oceanic crust is increasing in area by upwelling magma along the midocean ridges or world rift systems, and by moving away of the new material at rates of one to ten centimeters per year. This movement provides the source of dynamic thrust in the hypothesis of plate tectonics (AGI, 1997).

Sill. A tabular igneous body that parallels the bedding or foliation of the sedimentary or metamorphic country rock, respectively (AGI, 1997).

Sonoma Orogeny. An orogeny that affected northern Nevada and is typified by the pronounced unconformity between the little-deformed Lower Triassic Koipato group and the underlying, strongly folded, and thrust-faulted Mississippian to Permian oceanic rocks of the Golconda allochthon. Major eastward displacement of the Golconda allochthon on the Golconda thrust fault is inferred to have taken place during this orogeny (Silberling and Roberts, 1962).

Sphalerite. Zinc sulfide, weathers to white zinc carbonate or zinc sulfate.

Stock. A small intrusive body of igneous rock.

Syncline. A fold of which the core contains stratigraphically younger rocks; it is generally concave upward (AGI, 1997).

Thrust fault. A fault commonly of low dip in which the upper plate generally consists of older rocks that override younger rocks. Such faults may have displacements on the order of tens of miles (e.g., Roberts Mountains thrust).

Unconformity. A substantial break or gap in the geologic record where a rock unit is overlain by another that is not next in stratigraphic succession (modified from AGI, 1997).

Vein. An epigenetic mineral filling of a fault or fracture in a host rock, in tabular or sheetlike form (AGI, 1997).

Window. An eroded area of a thrust sheet that displays the rock beneath the thrust sheet (AGI, 1997).

References

Adkins, A. R., and J. C. Rota. 1984. General geology of the Carlin Gold Mine. In *Exploration for ore deposits of the North American cordillera, field trip guidebook,* ed. J. L. Johnson. Association of Exploration Geochemists Regional Symposium, March 1984, pp. FT1/17–FT1/23. Reno, Nevada.

Agricola, G. 1556. *De re metallica.* Trans. H. C. Hoover and L. H. Hoover. New York: Dover Publishing.

Al-Wohaibi, Abdullah. 1973. *The northern Hijaz in the writings of the Arab geographers, 800–1150.* Beirut: Al-Risalah Ets. 486 pp.

Armstrong, R. L. 1968. Sevier orogenic belt in Nevada and Utah. *Geol. Soc. Am. Bull.* 79:429–58.

Babcock, R. C., Jr., G. H. Ballantyne, and C. H. Phillips. 1995. Summary of the geology of the Bingham mining district, Utah. In Cordilleran Symposium, Porphyry copper deposits, from Alaska to Chile, October 5–7, 1994, pp. 316–35. Tucson, Arizona.

Bagdady, A. Y., J. W. Whitlow, and R. J. Roberts. 1979. Placer gold deposits in the northeastern part of Mahd adh Dhahab, Kingdom of Saudi Arabia. U.S. Geological Survey Saudi Arabian Proj. Rept. 235, 38 pp.

Baker, A. A. 1947. Stratigraphy of the Wasatch Mountains, Utah. U.S. Geological Survey, Oil and Gas Chart 30.

Bennett, W. G. 1937. Stratigraphic and structural studies in the Colville quadrangle. Ph.D. dissertation, University of Chicago.

Bibby, Geoffrey. 1969. *Looking for dilmun.* New York: Alfred A. Knopf.

Birak, D. J., and R. B. Hawkins. 1986. The geology of the Enfield Bell Mine and Jerritt Canyon district, Elko County, Nevada. In *Sediment-hosted precious metal deposits of northern Nevada,* ed. J. V. Tingley and H. F. Bonham. Reno: Nevada Bureau of Mines and Geology.

Bonham, H. F., Jr. 1982. Reserves, host rocks, and ages of bulk minable precious metal deposits in Nevada. Reno: Nevada Bureau of Mines and Geology, Open File.

Boutwell, J. M. 1905. Economic geology of the Bingham mining district, Utah. U.S. Geological Survey Prof. Paper 38, 413 pp.

Brew, D. A., and M. Gordon Jr. 1971. Mississippian stratigraphy of the Diamond Peak area, Eureka County, Nevada. U.S. Geological Survey Prof. Paper 661, 81 pp.

Brown, G. F., D. L. Schmidt, and A. C. Huffman. 1989. Geology of the Arabian Peninsula. U.S. Geological Survey Prof. Paper 560-A, 188 pp.

Burton, R. F. 1878. *The gold mines of Midian and the ruined Midianite cities.* London: C. Kegan Paul.

Capps, S. R. 1939. The Dixie placer district, Idaho, with notes on the lode mines by Ralph J. Roberts. Idaho Bureau of Mines and Geology Pamphlet 48, 45 pp.

Carlisle, D., M. A. Murphy, C. A. Nelson, and E. L. Winterer. 1957. Devonian stratigraphy of Sulphur Springs and Pinyon Ranges, Nevada. *Bull. Amer. Assoc. Petrol. Geol.* 41:2175–90.

Carter, K. B. 1947. *Pony Express, special edition.* Mid-Century Memorial Commission of Utah.

Clark, L. D. 1964. Stratigraphy and structure of part of the western Sierra Nevada metamorphic belt, California. U.S. Geological Survey Prof. Paper 410, 70 pp.

Coats, R. R., and J. F. Riva. 1983. Overlapping overthrust belts of late Paleozoic and Mesozoic age, northern Elko County, Nevada. In *Tectonic and stratigraphic studies in the eastern Great Basin,* ed. D. M. Miller et al., 305–27. Geol. Soc. Am. Memoir 157.

Coney, P. J. 1970. The geotectonic cycle and the new global tectonics. *Geol. Soc. Am. Bull.* 81:739–47.

Crawfurd, C. E. V. 1929. *Treasure of Ophir.* London: S. Keffington and Son.

Crittenden, M. D., Jr. 1959. Mississippian stratigraphy of the central Wasatch and western Uinta Mountains, Utah: Intermtn. Assoc. Petroleum Geologists, Guidebook, 10th Ann. Field Conf., 1959, pp. 63–74.

———. 1969. Interaction between Sevier Orogenic Belt and Uinta structures near Salt Lake City, Utah (abs.). *Geol. Soc. Am. Abst. with Programs* pt. 5, p. 18.

Davidson, M. B., ed. 1962. *Lost Worlds.* New York: American Heritage.

Dekalb, C. 1914. Mines and miners of Bible times. *Mining Mag.* 10, 5:362–69.

Dever, W. G. 1999. Histories and Nonhistories of Ancient Israel. *Bull. Am. Schools of Oriental Research* 316:89–105.

Dewey, J. F., and J. M. Bird. 1970. Mountain belts and the new global tectonics. *Jour. Ceophys. Research* 75:2625–47.

Dolph, O. P. 1942. King Solomon's Mine Arabia. *Mines Mag.* 32:21–24.

Eardley, A. J. 1934. The paleogeography of the southern Wasatch mountains near Provo, Utah. *Mich. Ac. Sci. Paper* 19:377–400.

Erickson, R. L., E. H. Van Sickle, H. M. Nakagawa, J. H. McCarthy Jr., and K. W. Long. 1966. Gold geochemical anomaly in the Cortez district, Nevada. U.S. Geological Survey Circ. 534, 9 pp.

Evans, J. G., and T. G. Theodore. 1978. Deformation of the Roberts Mountains allochthon in north-central Nevada: U.S. Geological Survey Prof. Paper 1060, 18 pp.

Fage, J. D., and R. A. Oliver, eds. 1972. *Papers in African prehistory.* Cambridge: Cambridge University Press.

Ferguson, H. G. 1924. Geology and ore deposits of the Manhattan district, Nevada. *U.S. Geol. Survey Bull.* 723, 163 pp.

———. 1929. The mining districts of Nevada. *Econ. Geol.* 24:115–48.

———. 1932. The geology and ore deposits of the Allegheny district, California. U.S. Geological Survey Prof. Paper 172, 126 pp.

———. 1940. The Mogollon mining district, New Mexico. *U.S. Geol. Survey Bull.* 787, 100 pp.

———. 1951a. Geologic map of the Winnemuca quadrangle, Nevada. U.S. Geological Survey Quad. Map GQ-11.

———. 1951b. Geologic map of the Mt. Moses quadrangle, Nevada. U.S. Geological Survey Quad. Map GQ-12.

Ferguson, H. G., R. J. Roberts, and W. W. Muller. 1952. Geologic map of the Golconda quadrangle, Nevada. U.S. Geological Survey Quad. Map GQ-15.

Forbes, R. J. 1964. *Studies in ancient technology.* New York: E. J. Brill.

Geob, I. J. 1970. Makkan and Meluhha in early Mesopotamian sources. *Revue D'Assyriologie* 64, 1.

Gilluly, J. 1932. Geology and ore deposits of the Stockton and Fairfield quadrangles, Utah. U.S. Geological Survey Prof. Paper 173, 171 pp.

———. 1968. Memorial to Joseph Hoover Mackin. *Proceedings of the Geological Society of America* (1968): 206.

Gilluly, J., and O. Gates. 1965. Tectonic and igneous geology of the northern Shoshone Range, Nevada, with a section on gravity in Crescent Valley, by Donald Plouff, and economic geology, by K. B. Ketner. U.S. Geological Survey Prof. Paper 465, 153 pp.

Gilluly, J., and H. Masursky. 1965. Geology of the Cortez quadrangle, Nevada, with a section on gravity and aeromagnetic surveys by D. R. Mabey. *U.S. Geol. Survey Bull.* 1175, 117 pp.

Goldsmith, R. 1971. Mineral resources of the southern Hijaz quadrangle, Kingdom of Saudi Arabia. *Saudi Arabian Dir. Gen. Mineral Resources Bull.* 5, 62 pp.

Goldsmith, R., and J. H. Kouther. 1971. Geology of the Mahd adh Dhahab-Umm ad Damar area, Kingdom of Saudi Arabia. *Saudi Arabian Dir. Gen. Mineral Resources Bull.* 6, 20 pp.

Hague, A. 1892. Geology of the Eureka District, Nevada. U.S. Geological Survey Mon. 20, 419 pp.

Hardie, B. S. 1966. Carlin Gold Mine Lynn district, Nevada. Nevada Bur. Mines Rept. 13, pt. A, pp. 73–83.

Hausen, D. M., and P. F. Kerr. 1968. Fine gold occurrence at Carlin, Nevada. In *Ore deposits of the United States, 1933–1967; the Graton-Sales volume,* ed. J. D. Ridge, 910–40. New York: American Institute of Mining, Metallurgical, and Petroleum Engineers.

Herzog, Z. 1999. Deconstructuring the walls of Jericho. *Ha'aretz Mag.* (October 29, 1999).

Hill, J. M. 1915. Some mining districts in northeastern California and northwestern Nevada. *U.S. Geol. Survey Bull.* 594, 200 pp.

Hilpert, L. S., R. J. Roberts, and G. A. Dirom. 1982. Geology of Mine Hill and the underground workings, Mahd adh Dhahab mine, Kingdom of Saudi Arabia. U.S. Geological Survey Tech. Rept. TR-03-2 (IR 592), 36 pp.

Hose, R. K., and Z. K. Danes. 1973. Development of the late Mesozoic to early Cenozoic structures of the eastern Great Basin. In *Gravity and Tectonics,* ed. K. A. de Jong and R. Scholten, 271–86.

Hotz, P. E., and C. R. Willden. 1964. Geology and mineral deposits of the Osgood Mountains quadrangle, Humboldt County, Nevada. U.S. Geological Survey Prof. Paper 431, 128 pp.

Hunt, C. B. 1956. Cenozoic geology of the Colorado Plateau. U.S. Geological Survey Prof. Paper 279, 99 pp.

Johnson, J. G., and A. Pendergast. 1981. Timing and mode of emplacement of the Roberts Mountains allochthon, Antler Orogeny. *Geol. Soc. Am. Bull.* 92:648–58.

Joseph, N. L. Geologic map of the Colville 1:100,000 quadrangle, Washington. Division of Geology and Earth Resources, Open File Rept. 90-13, pp. 1–78.

Keane, A. H. 1901. *The gold of Ophir.* London: E. Stanford.

Keller, W. 1956. *The Bible as history.* Trans. William Neil. New York: William Morrow.

Kemp, J. Gros, Y., and J. Prian. 1982. Geologic map of the Mahd adh Dhahab quadrangle, sheet 23 E, Kingdom of Saudi Arabia: Ministry of Petroleum and Mineral Resources, map GM-64A.

Kennedy, G. C. 1959. The origin of the continents, mountain ranges, and ocean basins. *Am. Jour. Sci.* 47:491–504.

Kent, C. F. 1905. *Israel's historical and biographical narratives from the establishment of the Hebrew Kingdom to the end of the Maccabean struggle.* New York: Charles Scribner's Sons.

King, C. 1878. Systematic geology. U.S. Geological Expl. 40th Par. Rept., 1:1–803.

Kirk, E. 1933. The Eureka quartzite of the Great Basin region. *Amer. Jour. Sci.* 5th Ser. 26:27–43.

Kitchen, K. A. 1989. Where did Solomon's gold go? *Biblical Arch. Rev.* 15, 3:30–31, 34.

Larson, E., and J. Riva. Geologic map and sections of the Diamond Springs quadrangle, Nevada. Nevada Bureau of Mines and Geology Map 20.

Lehner, R. E., M. M. Bell, K. M. Tagg, and R. J. Roberts. 1961. Preliminary geologic map of Eureka County, Nevada. U.S. Geological Survey Min. Invest. Map Series, MIF 176.

Levy, T. E., R. B. Adams, and M. Najjar. 1999. *Early metallurgy and social evolution, Jabal Hamrat Fidan, [Jordan]. ACOSR (American Center of Oriental Research) Newsletter* 11:1–6.

Luce, R. W., A. Bagdady, and R. J. Roberts. 1976. Geology and ore deposits at the Mahd adh Dhahab mine, Kingdom of Saudi Arabia. U.S. Geological Survey Saudi Arabian Proj. Rept. 195, 34 pp.

Lutz, H. F. 1924. Money and loans in ancient Babylonia. *California Univ. Chronicles* 6, 2:125–44.

MacDonald, G. J. F. 1963. The deep structure of continents. *Rev. Geophysics* 1, 4:587–65.

Mackin, J. H. 1937. Erosional history of the Big Horn Basin, Wyoming. *Geol. Soc. Am. Bull.* 48, 6:813–93.

Madrid, R., R. J. Roberts, and W. C. Bagby. 2000. Gold belts in north-central Nevada, GSA Symposium.

Mattison, J. M. 1972. Ages of zircons of the northern Cascade Mountains, Washington. *Geol. Soc. Am. Bull.* 83:3769–83.

Merriam, C. W., and C. A. Anderson. 1942. Reconnaissance survey of the Roberts Mountains, Nevada. *Geol. Soc. Am. Bull.* 53:1675–726.

A miner's dream. *Newsweek* (June 7, 1965):71–72.

Misch, P., and J. C. Hazzard. 1962. Stratigraphy and metamorphism of late Precambrian rocks in central northeastern Nevada and adjacent Utah. *Am. Assoc. Petrol. Geol. Bull.* 46:289–343.

Mullens, T. E. 1980. Stratigraphy, petrology, and some fossil data of the Roberts Mountains Formation, north-central Nevada. U.S. Geological Survey Prof. Paper 1063, 67 pp.

Muller, S. W., H. G. Ferguson, and R. J. Roberts. 1951. Geologic map of the Mt. Tobin quadrangle, Nevada. U.S. Geological Survey Quad. Map GQ-7.

Murphy, M. A., J. D. Power, and J. G. Johnson. 1984. Evidence for late Devonian movement within the Roberts Mountains allochthon, Roberts Mountains, Nevada. *Geology* 12:20–23.

Niebuhr, G. T. 2000. The Bible as history flunks new archeological tests. *New York Times* (July 29, 2000).

Nolan, T. B. 1928. A late Paleozoic positive area in Nevada. *Amer. Jour. Sci.* ser. 5, 16:153–61.

———. 1935. The Gold Hill mining district, Utah. U.S. Geological Survey Prof. Paper 177.

———. 1936. The Tuscarora mining district, Nevada. *Nevada Univ. Bull.* 30, 38 pp.

———. 1943. The Basin and Range province in Utah, Nevada, and California. U.S. Geological Survey Prof. Paper 2.

———. 1962. The Eureka mining district, Nevada. U.S. Geological Survey Prof. Paper 406, 78 pp.

———. 1974. Stratigraphic evidence on the age of the Roberts Mountains thrust, Eureka and White Pine Counties, Nevada. *U.S. Geol. Survey Jour. Res.* 2:411–16.

Nolan, T. B., C. W. Merriam, and J. S. Williams. 1956. The stratigraphic section in the vicinity of Eureka, Nevada. U.S. Geological Survey Prof. Paper 276, 77 pp.

Overstreet, W. C., D. J. Grimes, J. F. Seitz, and T. Botinelly. 1983. Ores and slags of copper and iron from ancient mines and smelters in the Hashemite Kingdom of Jordan. Field Research Projects, Miami, Fla., 46 pp.

Peters, K. 1902. The Eldorado of the ancients. London: Pearson, Ltd.

Poole, F. G. 1974. Flysch deposits of the Antler foreland basin, western United States. In *Tectonics and sedimentation,* ed. W. R. Dickinson, 58-82. Society of Economic Paleontologists and Mineralogists Spec. Pub. 22.

Presnell, R. D., and W. T. Perry. 1996. Geology and geochemistry of the Barneys Canyon gold deposit. *Econ. Geol.* 91:277–88.

Pritchard, J. B., ed. 1974. *Solomon and Sheba.* London: Phaidon.

Radtke, A. S. 1981. Geology of the Carlin gold deposit, Nevada. U.S. Geological Survey Prof. Paper.

Radtke, A. S., and B. J. Scheiner. 1970. Studies of hydrothermal gold deposition (I). Carlin gold deposit, Nevada: The role of carbonaceous materials in gold deposition. *Econ. Geol.* 65:87–102.

Rickard, T. A. 1932. *Man and metals.* New York: McGraw-Hill.

Riva, J. 1970. Thrusted Paleozoic rocks in the northern and central HD Range, northeastern Nevada. *Geol. Soc. Am. Bull.* 81:2689–716.

Roberts, R. J. 1945. Manganese deposits in Costa Rica. *U.S. Geol. Survey Bull.* 935-H:387–414.

———. 1949. Geology of the Antler Peak quadrangle, Nevada. U.S. Geological Survey Open File Rept. 108 pp.

———. 1951. Geology of the Antler Peak quadrangle, Nevada. U.S. Geological Survey Quad. Map GQ-10.

———. 1957. Major mineral belts in Nevada. American Institute of Mining

Metallurgy and Petroleum Engineers, Pacific Southwest Mineral Industry Confernce, Reno, Nevada. Abstracts of papers.

———. 1960. Alinement of mining districts in north-central Nevada, U.S. Geological Survey research 1960. U.S. Geological Survey Prof. Paper 400-B, pp. B17–BL9.

———. 1964. Stratigraphy and structure of the Antler Peak quadrangle, Humboldt and Lander Counties, Nevada. U.S. Geological Survey Prof. Paper 459-A, pp. A1-A93.

———. 1966. Metallagenic provinces and mineral belts in Nevada. Nevada Bureau of Mines Rept. 13, pt. A, pp. 47–72.

———. 1968. Tectonic framework of the Great Basin. In *A coast to coast tectonic survey of the United States. UMR Jour.* 1:101–19.

———. 1972. Evolution of the Cordilleran fold belt. *Geol. Soc. Am. Bull.* 83:1989–2004.

———. 1976. Mahd adh Dhahab—The Ophir of antiquity. Typescript. U.S. Geological Survey, Jiddah, Saudi Arabia.

———. 1991. Mahd adh Dhahab—The Ophir of antiquity? Arabia Antiqua Conference—The Mineral Wealth of Ancient Arabia, Ismeo-Instituto Italiano, Rome.

———. 1997. Carlin-type gold deposits: Their localization and genesis. Paper presented at Arizona Geological Society, March 1997.

Roberts, R. J., E. Sampson, M. M. Striker, and R. E. Stewart. 1949. Performance of the SCR–625 mine detector over different rocks and soils. Fort Belvoir, Va.: Engineering Research and Development Lab.

Roberts, R. J., and R. E. Lehner. 1955. Additional data on the age and extent of the Roberts Mountains thrust fault, north central Nevada. *Geol. Soc. Am. Bull.* 66:1661.

Roberts, R. J., and E. M. Irving. 1957. Mineral deposits of Central America with a section on manganese deposits of Panama by F. S. Simons. *U.S. Geol. Survey Bull.* 1034, 205 pp.

Roberts, R. J., P. E. Holtz, J. Gilluly, and H. G. Ferguson. 1958. Paleozoic rocks of north-central Nevada. *Am. Assoc. Pet. Geol. Bull.* 42, 12:2813–59.

Roberts, R. J., and E. W. Tooker. 1961. Structure of the north end of the Oquirrh Mountains, Utah. In *Geology of the Bingham mining district: Guidebook to the geology of Utah, Utah Geol. Soc.* 16:36–48.

Roberts, R. J., and D. C. Arnold. 1965. Ore deposits of the Antler Peak quadrangle, Humboldt and Lander Counties, Nevada. U.S. Geological Survey Prof. Paper 459-B, pp. B1-B94.

Roberts, R, J., M. D. Crittenden Jr., E. W. Tooker, H. T. Morris, R. K. Hose, and T. M. Cheney. 1965. Pennsylvanian and Permian basins in northwestern Utah, northeastern Nevada, and south-central Idaho. *Am. Assoc. Petrol. Geol. Bull.* 49:1926–56.

Roberts, R. J., K. B. Ketner, and A. S. Radtke. 1967. Geological environment of gold depoits in Nevada. *Applied Earth Sci. Trans. Inst. Min. Metall.* 76:78.

Roberts, R. J., and K. M. Montgomery. 1967. Geology and mineral resources of Eureka County, Nevada: *Nevada Bur. Mines Bull.* 64, 152 pp.

Roberts, R. J., A. S. Radtke, and R. R. Coats. 1971. Gold-bearing deposits in north-central Nevada and southwestern Idaho, with a section on periods of plutonium in north-central Nevada by M. L. Silberman and E. H. McKee. *Econ. Geol.* 66:14–33.

Roberts, R. J., and M. D. Crittenden Jr. 1973. Orogenic mechanisms, Sevier orogenic belt in Nevada and Utah. In *Gravity and Tectonics,* ed. K. A. de Jong and R. Scholten, 253–70.

Roberts, R. J., A. Bagdady, and R. W. Luce. 1976. Geochemical investigations in the Mahd adh Dhahab district, Kingdom of Saudi Arabia. U.S. Geological Survey Saudi Arabian Proj. Rept. 195, 29 pp.

Scott, R. B. Y. 1959. Weights and measures of the Bible. *Biblical Archeologist* 22, 2:22–40.

Seabrook, J. 1989. Invisible gold. *New Yorker* (April 24, 1989): 45 et. seq.

Silberling, N. J., and R. J. Roberts. 1962. Pre-Tertiary stratigraphy and structure of northwestern Nevada. Geology Society of America Spec. Paper 72, 58 pp.

Skipp, B., W. J. Sando, and W. E. Hall. 1979. The Mississippian and Pennsylvanian (carboniferous) systems in the United States—Idaho. U. S. Geological Survey Professional Paper 1110, pp. AA1–AA41.

Smith, J. F., and K. B. Ketner. 1968. Devonian and Mississippian rocks and the date of the Roberts Mountains thrust in the Carlin-Pinon Range area, Nevada. *U.S. Geol. Survey Bull.* 1251, 1:1–19.

Speed, R. C., and N. H. Sleep. 1982. Antler Orogeny and foreland basin—a model. *Geol. Soc. Am. Bull.* 93:815–28.

Spieker, E. M. 1946. Late Mesozoic and early Cenozoic history of central Utah. U.S. Geological Survey Prof. Paper 205-D, pp. D117–D161.

Stewart, J. H. 1980. *Geology of Nevada.* Nevada Bureau of Mines and Geology Special Publication 4.

Stille, H. 1940. *Einfuhrung in den Bau Americas* Berlin: Borntraeger.

Theodore, T. G., M. L. Silberman, and D. Blake. 1973. Geochemistry and K-Ar ages of plutonic rocks in the Battle Mountain mining district, Lander County, Nevada. U.S. Geological Survey Prof. Paper 798-A, 24 pp.

Tooker, E. W., and R. J. Roberts. 1970. Upper Paleozoic rocks in the Oquirrh Mountains and Bingham mining district, Utah. U.S. Geological Survey Prof. Paper 62A, pp. A1–A76.

———. 1992. Geology of the Oquirrh Mountains, Utah. U.S. Geological Sut. Prof. Paper 629-C, 139 pp.

———. 1996. Sevier-age thrust fault structures control the location of base- and precious-metal mining districts in the Oquirrh Mountains, Utah. U.S. Geological Surv. Paper 629-D, 43 pp.

———. 1999. Geologic map of the Oquirrh Mountains, Utah. In preparation.

Twitchell, K. S. 1958a. *Saudi Arabia, with an account of the development of its natural resources.* 3d ed. Princeton: Princeton University Press.

———. 1958b. Cradle of gold. *Explorer's Jour.* 36, 4:14–21.

Vanderberg, W. O. 1938a. Reconnaissance of mining districts in Eureka County, Nevada. U.S. Bureau of Mines Inf. Circ. 7022, 66 pp.

———. 1938b. Reconnaissance of mining districts in Lander County, Nevada. U.S. Bureau of Mines Inf. Circ. 7043, 83 pp.

White, D. E. 1955. Thermal springs and epithermal ore deposits. *Econ. Geol.* Fiftieth anniversary volume, pt. 1.

Whitehill, H. R. 1873. Biennial report of the state mineralogist of the State of Nevada for the years 1871 and 1872. Carson City, Nevada.

Wilkie, P. J., L. A. Quintero, and G. O. Rollefson. *ACOR Newsletter* 10.2 (winter 1998):6–7.

Worl, R. W. 1978. Mineral exploration: Mahd adh Dhahab district, Kingdom of Saudi Arabia. *SAIR* 233, Dir. Gen. of Petroleum and Mineral Resources, Kingdom of Saudi Arabia.

Wrucke, C. T., and T. J. Armbrustmacher. 1969. Structural controls of the gold deposit at the open-pit mine, Gold Acres, Lander County, Nevada. Geology Society of America Cordilleran Section Sixty-fifth Annual Meeting, Eugene, Oregon. Abstracts with Programs for 1969, pt. 3, p. 75.

Index